# DEEP INFORMATION:
## The Role of Information Policy in Environmental Sustainability

# Information Management, Policy, and Services

........................

## Peter Hernon, series editor

# DEEP INFORMATION:
## The Role of Information Policy in Environmental Sustainability

John Felleman
State University of New York, Syracuse

Ablex Publishing Corporation
Greenwich, Connecticut
London, England

Printed in the United States of America

Library of Congress Cataloging-in-Publication Data

Felleman, John.
    Deep information : the role of information policy in environmental
sustainability / John Felleman.
        p.   cm. — (Information management, policy, and services)
    Includes bibliographical references and index.
    ISBN 1-56750-303-9 (cloth).  —  ISBN 1-56750-304-7 (pbk.)
        1. Environmental management—United States. 2. Sustainable
development—United States. 3. Information policy—United States.
    I. Title. II. Series
    GE300.F45   1997
    363.7'00685—dc21                                         96–52275
                                                              CIP

Ablex Publishing Corporation            Published in the U.K. and Europe by:
P.O. Box 5297                           JAI Press Ltd.
55 Old post Road #2                     38 Tavistock Street
Greenwich, CT 06831                     Covent Garden
                                        London WC2E 7PB
                                        England

# Contents

........................

# Preface

........................

The origin of this book was a sabbatic leave, where as a Senior Environmental Fellow at the Rockefeller Institute of Government, I had planned to study the application of emerging database and geographic information system applications to the management of the New York state's recently designated federal sole source aquifers. It quickly became apparent that even though information technology was being rapidly applied to this and similar environmental problems, there were a number of distinct disciplinary subcultures working in parallel. The first and second generation information systems were reinforcing the traditional isolation of the land information, regulation and development, and scientific communities.

The simple exchange of data products from these independent systems was clearly inadequate for the challenge of protecting a resource such as a regional sole source aquifer in perpetuity. The aquifer case study was abandoned and a more fundamental examination was begun of the opportunities and barriers for an information environment that was up to the challenges of our physical environment in a pluralistic democracy. Out of this came the core concepts of "deep information".

Upon return to teaching, I replaced my graduate course in Landuse Information with one entitled Environmental Information Policy. Counsel was sought and warmly provided from Chuck McClure, whose Federal Information Policy course has been highly successful at Syracuse University. The new offering was specifically designed to have a broad spectrum of students from policy, science, engineering, and information systems backgrounds which reflected the outside "real world". The Greenetowne scenario was developed to ground the basic issues and to introduce the differing needs of the various information communities which have evolved legacy information systems.

In the past six years graduate students have applied the basic diagramming approaches and critical deep information objectives to over a hundred case studies ranging from global warming to backyard septic systems. Their feedback has been invaluable and a number of the examples used here reflect their research. Any errors in this presentation are solely the responsibility of the author. An article entitled "Deep Information" was published in the *URISA Journal* (1994, v.6,n.2,11-24), and the response and constructive criticisms from this effort led the fuller expansion of the topic into book form. Thanks to Ralph Sanders, who as Department Chair provided both encouragement and that critical commodity, schedule flexibility; and to Adam Felleman whose VISIO-4 computer graphics play a critical role in the effort.

Deep Information is intended for a broad audience of information, environmental, and policy persuasions. It has a clear bias: information is essential to environmental sustainability, and to be useful the systems we develop must deeply reflect the environment, capitalism, and democracy. The book was delayed a bit as we all watched the rapid evolution of the administration's reinventing government and the "contract" of the new republican congress. Although some of the examples used in this volume may have a short life due to these actors, both the administrative and congressional processes can be interpreted as a fundamental shift in our policy framework, and a ripe opportunity for a full dialog on deep information.

**John Felleman**
**Syracuse**
**June 1996**

# 1

## Introduction

### INFORMATION AND DEMOCRACY

*T*he present environmental era emerged as a central component of the sweeping post-World War II social changes that crystallized in the turbulent 1960s. Its initial benchmark, the National Environmental Policy Act (NEPA 1970, 42 USC 4321 et seq.), contains a fascinating, internally contradictory view of the larger societal shifts. NEPA is essentially an information policy, with the environmental impact statement (EIS), which must precede every major Federal environmentally related decision, as its keystone. In the main, the Act embraced the prevailing logical positivism of the modern era. Decisions were to be based on the "systematic, interdisciplinary... use of the natural and social sciences" (Sec. 4332 (A)) to predict the future impacts of proposed actions. Where such sciences were currently lacking, the Federal Government was to:

> identify and develop methods and procedures... which will insure that presently unquantified amenities and values may be given appropriate consideration in decision making alongside economic and technical considerations. (Sec. 4332 (B))

This approach reinforces the prevailing scope and influence of the technocracy.

From an alternate perspective, NEPA can be viewed as a partial rejection of the centralized, closed door, science- and technical-based elitism that had dominated the twentieth century. Drawing in part on the widely publicized efforts of Rachel Carson to expose the effects of broad scale use of chemicals, and the deepening opposition to nuclear power, and interstate highway and dam construction, NEPA strongly reengaged the historical continuum of pluralism and populism that runs directly back to Jeffersonian democracy. Information regarding the environment is to be made "available to States, counties, institutions, and individuals" (Sec. 4332 (G)). The draft EIS is open for comment to other Federal agencies, and affected State and local governments and tribes. The final EIS must include "affirmatively solicited comments from those persons or organizations who may be interested or affected" (CFR 40 Ch V Council on Environmental Quality Sec. 1503.1). Such comments are to assessed, considered, and responded to (Sec. 1503.4).

The elitist versus populist dualism embedded in NEPA stretches across the entire information spectrum, from data generation, to information analysis, to the development of predictive knowledge, and finally to the wisdom of decision in the light of incomplete information and knowledge. Its roots lie at the foundation of eighteenth-century enlightenment, which embraced both the objectivity of science and the virtue of the public. From a policy standpoint, the environment is a relatively recent focal arena, but the underlying tension has deep roots. By the 1920s, the rise of large professionally-staffed governmental bureaucracies had emerged as a topic of major concern. For over a decade, two leading American figures debated the fate of participatory democracy in a modern industrial economy. Walter Lippmann, the most widely respected journalist of the era, concluded:

> In the absence of institutions and education by which the environment is so successfully reported that the realities of public life stand out sharply against self-centered opinion, the common interests very largely elude public opinion entirely, and can be managed only by a specialized class where personal interests reach beyond the locality. (Westbrook, 1990, p. 298)

Lippmann held that Jefferson's agrarian ideal of the informed citizenry was not feasible in a highly complicated society whose main channel of information was the newspaper.

Championing the populist position was John Dewey, the philosopher and educational reformer. He was convinced that, through education, citizens were capable of understanding scientific issues, and that by utilizing reflective thinking and substantive dialog they could resolve complex problems. He did agree with Lippmann that modern society had overwhelmed traditional communities:

> Indirect, extensive, enduring and serious consequences of conjoint and interacting behavior call a public into existence having a common interest in controlling these consequences. But the machine age has so enormously expanded, multiplied, intensified and complicated the scope of indirect consequences, has formed such immense and consolidated unions in action, on an impersonal rather than a community basis, that the resultant public cannot identify and distinguish itself. And this discovery is obviously an antecedent condition of any effective organization on its part. (Dewey, p. 307)

Dewey called for the reconstruction of local communities that would socially develop policy through open discussion and debate. In particular, he argued that elites had different value sets than community publics, and therefore shouldn't be the prime decision makers. He did not,however, articulate how such communities were to be informed. His outlook on this topic was not optimistic.

Between the debate of the 1920s and the 1960s which spawned NEPA there were major transformations in the mass media (magazines, radio, TV), and in the science and environmental education and literacy of the citizenry. These were facilitated by major new information policies, such as the Administrative Procedures Act (APA) in 1946 (5 *USC* 551), the Freedom of Information Act (FOIA) in 1966 (5 *USC* 552 et seq.), and the establishment of the National Technical Information Service in 1970. Collectively, these policies greatly enhanced the ability of individuals, corporations, and interest groups to identify and access the explosively increasing amount of government environmental information. The ability to develop "public judgement" on complex issues such as fluoridation and nuclear power demonstrates an emerging capacity for the construction of interest communities, and the effectiveness of participatory democracy (Yankelovich, 1991).

This period can be seen as an extension of the modern program. Lippman, Dewey, and NEPA all basically view decision making as a rational two-phase process: objective science-based predictive analysis of alternatives followed by value-based administrative and/or political conclusions. Although the separation of "fact" and "values" had long been cast in doubt (Simon, 1976), the rational ideal remains a central foundation of modernism.

Paralleling the environmental movement of *Silent Spring*, Earth Day, Federal air and water regulations, and NEPA was the information revolution. The emerging availability of centralized mainframe computing in government and academia led in the mid-1960s to a decade of heady projections by natural and social scientists that computer modelling would provide the basic means to solve heretofore overly complex process analysis problems. Water quality models were required as a basis for approving waste water treatment grants; land use and traffic models were required for approving metropolitan interstate highway construction. This positive attitude toward modelling provided a major elitist influence for the construction of the predictive mandates in NEPA.

The period following the passage of NEPA was initially characterized by a continuation of positivism, the implementation of numerous top-down, command and control regulatory programs for air, water, and solid waste discharges. The primary focus of these controls were major stack and pipe, industrial and municipal "point" source locations. Significant pollution reductions were achieved. Program aid was initially provided to municipalities, and tax incentives were given to industry to facilitate the adoption of newer technologies. These programs and their accompanying impact statements were highly reductionist. The typical focus was a single or small set of criteria such as carbon monoxide in air, dissolved oxygen in water, or channel capacity for stream runoff. To oversee these regulatory and aid programs, environmental agencies utilized newly available database technologies to develop management information systems (MIS). The data stored were primarily selected to the meet legal procedural requirements of the technocracy. The vast majority of environmental description and analysis remained in hard-copy form. The digital and hard-copy information were often inaccessible except through a tedious FOIA request process.

Over the last decade, the effectiveness of these regulatory management approaches has become widely challenged. Environmental degradation, such as the filling of wetlands, slowed, but the trend continued. Growing criticism has focussed on the red tape and inefficiency of the regulatory process. It was revealed that a large fraction of the resources available for Superfund cleanup was spent on litigation instead of site mitigation. The Environmental Protection Agency (EPA) itself concluded:

> Because EPA has concentrated on issuing permits, establishing pollution limits, and setting national standards, the Agency has not paid enough attention to the overall environmental health of specific ecosystems. In short, EPA has been "program driven" rather than "place driven". Recently, we have realized that, even if we had perfect compliance with all our authorities, we could not assure the reversal of disturbing environmental trends. (EPA Ecosystem Protection Workgroup, 1994, p.1)

The central reliance on permits for environmental management had institutionalized disjointed incrementalism in the form of thousands of project decisions isolated in space and time. The initial promise of mathematical impact modelling was tempered by the stark realization of its inherent limits. This is due to the complex, often chaotic nature of systems, the role of human willfulness in changing socio-economic processes such as energy consumption, and the vast data requirements due to "genius loci", the uniqueness of local places. The scientific predictive capability which is the underlying rational basis of the impact statement proved to be quite modest. A growing gap was seen between the prevailing decision reductionism and the obviously integrated wholeness of the environment, as exhibited by global warming, the ozone hole, and loss of species diversity and critical habitats. As point sources became less critical, attention began to shift to dispersed "nonpoint" sources, including homeowners, farmers, and internal manufacturing practices.

Paralleling this transformation of environmental understanding was the second phase of the information revolution, with its shift from centralized to distributed systems. Networks coupled with increasingly powerful personal computers and associated software have brought both data and analysis tools to remote desktop end users. Major information and environmental policy changes have dramatically reinforced this trend. Federal information policy regarding public access has evolved from the negative "need to know" of the original APA, to the neutral "right to know" of FOIA, to the proactive open systems electronic dissemination called for in Office of Management and Budget Circular No. A-130 (1985). Environmental information has been in the forefront of this latest transformation to open information. The Toxic Release Inventory and Community Right to Know legislation (Emergency Planning and Community Right to Know Act, EPCRA, Title III of the Superfund Amendment and Reauthorization Act, SARA, 1986 42 USC 11001 et seq.), have taken much of the previously hidden maze of transactions between regulators and industry and made it readily available in simple electronic formats.

Recent policies have created the Federal Geographic Data Committee (FGDC) and charged it with overseeing the development of the National Spatial Data Clearinghouse, which provides remote access to Federal and cooperating state, local, and private sector geographic information system (GIS) data (OMB Circular N. A-16,

1990, and Executive Order 12906, 1994). The Clearinghouse utilizes approved Metadata and Spatial Data Transfer Standards.

Lippmann and Dewey had chronicled the demise of effective citizens and communities at the start of the century due to the overwhelming complexity of industrialized society and the highly limited existing channels of information. It would have been impossible for them to imagine a local Non-Governmental Organization (NGO) in New Jersey with its own GIS and trained user, remotely downloading State and Federal environmental attribute and pollution emission files, and developing a custom model to assess the implications of a proposed controversial project (NJ Department of Environmental Protection, 1995).

## SUSTAINABILITY

While the rapid advances in information technology and policy have reinvigorated the informed pluralism which is essential for a viable democracy, it remains to be seen whether this democracy has the capacity to steward its environment. As Federal, State, and local development programs, and regulations of the private sector, have multiplied, and the environmental sciences expanded, the ability of citizens, interest groups, and business to understand and participate effectively in public decision making remain seriously challenged. A central thrust of public environmental information policy of the past two decades has been to increase the transfer of existing hard-copy tabular and mapped data to digital form, and to open this information to a broad set of internal and external end-users. This focus has governmental efficiency and accountability as its primary objectives.

As our awareness of the deep integration of human and natural systems has matured, however, the prevailing modern paradigm of environmental "management" as a complement to continual consumption-based industrial development has been called into question. Information about electrical consumption, travel, and crop production improves our ability to locate power plants, new highways, and aerial crop spraying. Remote sensing and global positioning systems increase our capacity to document the loss of wetlands and endangered species. Advanced continuous monitoring technologies allow the tracking of atmospheric and underground pollution plumes. The essence of the managerial approach to the environment is to stop projects which are clearly destructive, such as the infamous Everglades jetport, and to "ameliorate" the vast majority of development proposals to an "acceptable" level of local disturbance.

This approach coupled with the open exchange of information may satisfy the precepts of democracy, but it has become increasingly obvious that the cumulative effects of thousands of disjointed incremental decisions has dramatically affected all scales of environmental systems, from local habitats to world climates. Ad hoc information describing environmental attributes is not equivalent to knowledge about the health and viability of complex environmental systems.

In the 1980s, a philosophy quite opposite to the dominant environmental management approach emerged. It became known as "deep ecology." This approach placed humans as a component in a highly complex, holistic, unified environmental system. It emphasized the process flows within the environment and concluded that human activities were destroying the planet. From this perspective of interconnection, deep ecologists proposed radical "solutions", including severe population controls, and fundamental restructuring of industrialization, energy utilization, and mass consumerism.

Although deep ecology may appear too radical for widespread adoption, by articulating an antithesis to environmental management it has focussed debate and spawned a potential synthesis, "sustainable development." In recent years "sustainability" has replaced management as the major goal concept in the ongoing environmental debate. To sustain is to perpetuate over time. Most physical, biological,and social systems are both internally dynamic and open to flows from surrounding systems. They exhibit the ability to self-correct and adapt within some range of external influences. Beyond this range, they become unstable and may collapse or transform into quite different systems.

Although not as revolutionary as deep ecology, sustainable development is still highly controversial. Much of the debate involves differences in means and ends, what "system(s)" to sustain. Ecological sustainability focuses on natural systems. Economic sustainability focuses on managed economic growth. Cultural sustainability emphasizes the perpetuation of traditional human lifestyles (Berke & Kartez, 1995).

Many take the position that modern development, by definition, is incompatible with environmental or cultural sustainability. They point to the limited success of three decades of both governmental command and control pollution regulations and efforts to protect critical habitats as a sign that these have failed to address the underlying causes rooted in our production/consumption, growth-based economy. Others look at recent environmental history as phase one of a multi-step transition to a post-industrial economy. They differentiate between the initial focus on major "point" sources of pollution, exhaust stacks and pipe discharges, and the more pervasive challenge of "non-point" sources, such as agricultural runoff and residential on-site sewage disposal. The emergence of NGOs as key actors in community-based pollution reduction and habitat preservation is cited as an example of a major shift in public awareness and responsibility sharing.

Regardless of the ultimate ends, consensus regarding five major points has emerged. First, due to its dynamic complexity, the environment is best understood from a systems perspective. However, our knowledge of these dynamic processes is quite limited. Second, the long-term health of socio-economic systems is intrinsically linked to the health of the supporting natural environment. Third, due to industrialization and population growth, human activity now affects virtually all natural systems, locally, regionally, and globally. Fourth, effective environmental stewardship cannot be achieved solely through a regulatory command and control managerial dialog between government and industry. All citizens and landowners have an interconnected set of environmental rights and responsibilities. Fifth, information systems and associated policies will play a pivotal role in the transformation from a managerial to a sustainable paradigm.

Although a democracy of informed citizens is an important objective in its own right, it is an insufficient foundation for sustainability. It will be necessary for business and communities to move beyond adversarial, check-and-balance, short-term objectives, such as deregulation or "not in my back-yard" (NIMBY), to a shared perspective of proactive stewardship. This transformation of responsibility to the environment is the fundamental information challenge.

## INFORMATION PUMPS AND FILTERS

From a general systems perspective we can identify two primary systems, environmental and human, as shown in Figure 1.1. Flows within the environmental system and its numerous subsystems involve continuous exchanges of matter and energy, such as ocean currents, climate, vegetative cycles, and geologic change. Flows between the natural and human systems are also primarily physical, including deforestation, agriculture, fishing, urbanization, and industrial emissions. The human to environmental physical flows are supported by information-gathering flows, such as reconnaissance mapping and field sampling followed by laboratory analyses.

Within the human system, flows are both physical, such as manufacturing, trade, and energy transport, and informational. The latter include the formal organization of the society, such as laws; the formal and informal conduct of commerce; the humanistic activities including arts and religion; and the continuous interpersonal communications. Together, these information flows critically affect interactions with the environment, and strongly influence human attitudes and behavior. They in turn have been fundamentally framed by the basic bifurcation in the modern organization of our

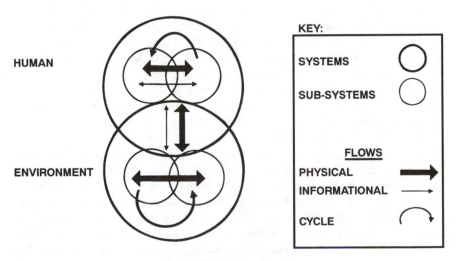

**Figure 1.1. Primary Systems**

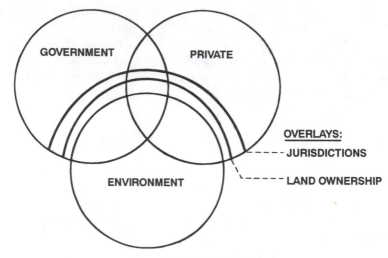

**Figure 1.2.   Social Subdivisions**

human activities, with its division between the governmental sector and the private sector. This is depicted in Figure 1.2.

Society's fundamental economic engine is capitalism, the private ownership of the means of production. The linchpin of a capitalist system is land ownership, the ability to own, use, and transfer land. Landowners, both private and public, have the ultimate responsibility for environmental stewardship. In Figure 1.2, land ownership is shown as a social "overlay" on the physical environment, consisting of property boundaries and associated rights, restrictions, and responsibilities. Pure unfettered capitalism is inherently unstable as a social system. It has historically led to dysfunctional markets and devastated environments. The development of strong, democratic governments has sought to maintain and enhance the creativity and productivity of capitalism while achieving three additional objectives: the provision of shared services, such as defense; the protection of public goods, such as clean air; and the enhancement of individual liberties. Government jurisdictions constitute the second social environmental overlay.

The result is a composite structure of physical flows and information flows, with a richly interwoven collage of environmental data. The cause of an information flow acts as a "pump" generating and transporting patterned data. Figure 1.3 illustrates these simplified relationships as generalized process flows. The private sector owns and operates facilities such as homes, farms, and factories (a). Activities of the governmental sector are commonly divided into two classes of programs, Provision and Regulatory. Provision programs include owning and operating physical activities such as schools, parks, military bases, and highways (b). A second set of Provision programs involve resource transfers to support the private sector, including monetary subsidies and information, such as census data and technology transfer (c).

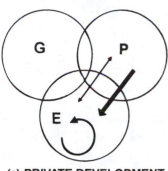

**(a) PRIVATE DEVELOPMENT**

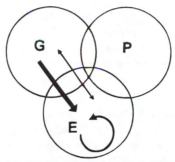

**(b) GOVERNMENT DEVELOPMENT**

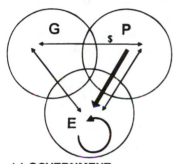

**(c) GOVERNMENT
RESOURCE TRANSFERS**

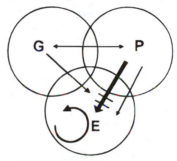

**(d) GOVERNMENT
ENVIRONMENTAL REGULATIONS**

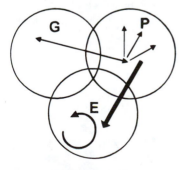

**(e) GOVERNMENT
INFORMATION REGULATIONS**

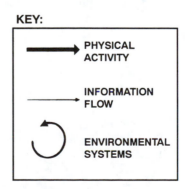

KEY:

PHYSICAL
ACTIVITY

INFORMATION
FLOW

ENVIRONMENTAL
SYSTEMS

Figure 1.3. Physical and Information Flows

9

In contrast, environmental regulatory programs are designed to place constraints and/or operating requirements on private and (some) governmental sector activities. Physical limits are typically placed on interfaces such as chemical air discharges (d). In addition, the intent of some recent regulatory programs is primarily informational, e.g. informing plant workers, neighbors, and local emergency staffs about the presence of toxic materials (e).

As shown in Figure 1.3, governmental programs, service and regulatory, are directly and indirectly major determinants of the character and quality of the physical environment. However, the flows of environmental information generated by the policy "pumps" are not entirely free. Positive and negative "filters" have been established to enhance, direct, and block selected transfers. Typically, governmental programs in a democracy are created by the legislature and implemented by the executive branch. Legislatures are invested with strong investigatory powers to develop and review programs. Executives have broad flexibility in developing, delivering, and monitoring programs. Democratic accountability is founded on the principle of an informed citizen. Citizens can elect chief executives and legislators who support particular views on the relationships of private sector activities and governmental programs. In addition, many recent environmental statutes allow citizens to sue governments who fail to implement the law. Accountability issues include timeliness, efficiency, effectiveness, and equity. Therefore, broad private sector access to governmental program information is essential for a viable system of checks and balances.

Figure 1.4 shows some of the fundamental information policy filter sets. Within government, there are internal administrative policies dealing with data standardization and intra-agency sharing. The primary objective of these is efficiency. Other policies filter the access and dissemination of government information to the private sector. Within the private sector, we can identify two basic subgroups, citizens and corporations. Information flows regarding the former are filtered by policies protecting privacy, while flows regarding the latter are filtered by policies protecting trade secrets and confidentiality.

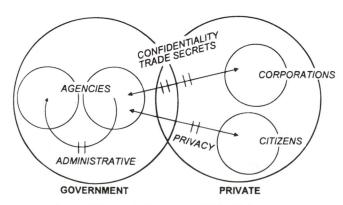

**Figure 1.4.  Information Policy Filters**

## ENVIRONMENTAL KNOWLEDGE

The physical environment integrates the interaction of ongoing natural processes and literally millions of private and governmental sector activities reflecting a wide spectrum of decision making processes. An assumption underlying any program of environmental sustainability is that decisions will be essentially rational. That is, they will incorporate an analysis of the expected outcomes of the proposed and alternative actions, and use this information explicitly in reaching the decision. This approach is the basis for the National Environmental Policy Act, and most subsequent Federal and State environmental laws.

A fundamental challenge to this form of logical positivism is immediately evident. Does society in general, and the government in particular, have the capacity to understand the effects of both current activities and proposed actions? This question confronts the linked continuum of environmental data, information, and causal knowledge. The modern managerial policy focus on information dissemination and pluralism presumes that the government has adequate descriptions of the status of the environment, and that government technocrats have the modelling capability to make adequate forecasts and predictions. The NEPA rational model is founded on this assumption.

Do we have adequate descriptive information on the current state of the environment, and adequate causal understanding of the process functioning of environmental systems? As described above, the drafters of NEPA were quite optimistic. Over the past two decades, most of the reform and adjustments to the impact process have been focused on paperwork reductions and efficiencies. Although the substantive quality of NEPA impact statements have not been well analyzed, it is widely accepted that the statements' predictions are frequently mediocre at best. In one of the few broad reviews, it was concluded that most impact predictions are very imprecise.

Probably the most important conclusion of this study was that, for most impact predictions, there was no available way to longitudinally check accuracy because there was typically no post-development monitoring (Culhane et al., 1985). This critical learning step is in fact optional in NEPA and its implementing regulations. Even for this optional monitoring, the intended focus is on checking that "mitigations" called for in the impact analysis were implemented, not on ascertaining whether the predictions were valid.

The weakness of our predictive capabilities is not limited to impact statements. A typical environmental regulation, whether for air, water, land, or ecosystem, is a political response to a perceived problem, often a "crisis" such as Bhopal, or Love Canal, or the loss of eagles, salmon, or other highly-valued species. A dearth of environmental knowledge is not necessarily a constraint on Congressional action. Implementing agencies are frequently in a position of developing regulations and procedures which balance the pragmatic resource concerns of the regulated community and the limited resources of the agency while striving to meet the often lofty goals of the legislation. This regulatory process has been frequently characterized as a house of cards built on a quicksand foundation of ignorance.

An example of this predictive dilemma can be seen in the handling of ecosystems. In the late 1960s ecology emerged, along with human health, as a primary focal arena for both popularized public awareness and much of the new academic interest in environmental science. Its underlying systems worldview is fundamental to the shift toward a sustainability paradigm. NEPA embraced this subject in a positivistic manner. Special emphasis was given to "initiate and utilize ecological information…" (Sec. 4332 (H)). Twenty years, millions of dollars, and numerous studies later, the perspective has become much more humble as the EPA abandons its grandiose plans for a national ecologically-based Environmental Monitoring and Assessment Program (EMAP).

Although these predictive problems may be news to legislatures and the public (in large part due to the self-promotion of scientific and technical communities), they are well understood. In a summary of mathematical modelling, Karplus took a comprehensive look at the information-knowledge continuum as shown in Figure 1.5 (Karplus, 1983). In a "white box" situation, such as the computer on which this is composed, we have both lawlike understanding of the underlying physical processes, and a high degree of descriptive information of the components due to the use of manufactured materials in a highly quality-controlled facility. In the "grey zone," which includes most of the natural environment, we have both an incomplete understanding of causality and major data limitations. Finally, Karplus relegates most human activities to the "black box" end of the spectrum. Here we encounter both major data problems and the complexity introduced by the willfulness and flexibility of human behavior.

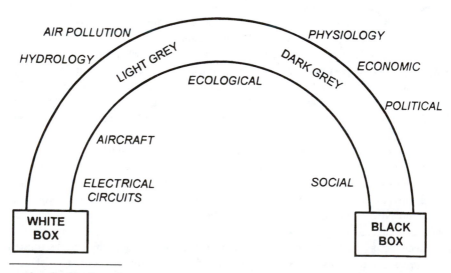

<sup></sup> a based on Karplus, p. 6

**Figure 1.5. Spectrum of Predictive Modeling**[a]

## DEEP INFORMATION

If descriptive information is incomplete and often obsolete, and causal knowledge is inherently limited, is the sustainability goal simply a mirage? In his important analysis of the troubled environment in the Pacific Northwest, Lee takes a cautiously optimistic view. His construct, "compass" and "gyroscope," is a process-oriented approach that addresses the critical middle range bridging question of "how do we get there from here?" Lee's overarching theme is social learning, a "combination of adaptive management and political change" (Lee 1993, p. 8). The "compass" is adaptive management, aggressively learning from the successes and failures of multiple decisions. The "gyroscope" is the bounded political process in which public and private, actors and stakeholders, conflict, debate, and negotiate in a context continuously informed about the environment.

The "reality check" of Karplus' spectrum in no way dooms the program of environmental sustainability. It is particularly useful in framing a policy agenda. First, we must be increasingly effective in utilizing the information and knowledge that are possessed in all sectors of the society. Next, we need to continue work to "lighten" those components of the spectrum that lend themselves to scientific study. Finally, we need to deal candidly with our limited analytical capabilities, including uncertainty and risk, and develop wise decision strategies that transcend formal rationalism in protecting environmental systems.

Comprehensive approaches to sustainability necessarily must combine continued environmental knowledge and the building of phenomena of our mixed private capitalism/governmental program economy, within the context of participatory democracy. The primary means of orchestrating these components is informational. Guidance of effective environmental information generation, dissemination, and stewardship is the role of public policy.

Numerous information-oriented policies currently exist. Many deal with the basic interfaces between governmental and private sectors, some deal with standardization of bureaucratic operations, while others address the myriad individual environmentally-related programs which have emerged in recent years. All reflect the information technologies that were dominant at a particular time. Cumulatively, these individual policies represent a crazy quilt of complements, contradictions, overlaps, and gaps that constitute a set of opportunities and barriers to environmental sustainability. In contrast, the accelerated development of electronic information systems over the last decade has emerged as a robust networked entity. The technology is now far ahead of the policy.

"Deep Information" represents a comprehensive meta-policy approach intended to facilitate environmental sustainability by enfolding the integrated analytical power of our new information systems. It builds on the basic premise of deep ecology, the interconnectedness of all human and natural systems. Simply stated, the goal of deep information policy is to facilitate action-oriented, effective societal learning. Deep information has three interrelated components, each reflecting a dominant American sustainability "reality": environmental systems ignorance ("compass"); democratic pluralism ("gyroscope"); and a capitalistic economy (Lee's navigational construct failed to articulate the vehicle's "engine"). The components are:

1. Shared, continuous knowledge-building of hierarchal environmental processes and the impacts of human activities;
2. Open, plural information generation, analysis, and dissemination; and
3. Downstream end-user orientation with particular emphasis on regional and local governments, communities, and landowners/facility managers.

## ORGANIZATION OF THE BOOK

The Introduction has utilized a highly compressed and obviously selective historical argument to track the interwoven evolutionary paths of environmental change, environmental world views, and information technology and policy. Flow diagrams have provided an important means for engaging environmental and informational system concepts. The next two chapters in Section 1 will expand on these themes. Chapter 2 includes a more detailed development of conceptual data flow diagramming which will then be used throughout the book.

Because the concepts of environment, information, and policy are so inclusive, it is important to ground the discussion. In Chapter 3. we visit a hypothetical region, Greenetowne, and examine its sustainability history through the lens of information. Chapter 4 is a brief overview of environmental data, information, and systems knowledge.

Although a new meta-policy approach such as Deep Information may be seen in part as a response to previous policies, it is important to understand that no new meta-policy can wipe the slate clean. Any chance for a successful transformation to Deep Information must include an understanding and integration of legacy information systems and their policy contexts. Section 2 is a critical examination of our primary environmental information legacy systems. First, Chapter 5 introduces the basic information policy frameworks of generation and analysis, dissemination, and stewardship. Chapter 6 covers the land ownership cadastre and its partial evolution to land information systems (LIS). Chapter 7 focuses on government programs, particularly environmental regulations, with their emphasis on management information systems (MIS). Chapter 8 examines the process of information as self regulation. Chapter 9 examines science information and its complex transformation from a reductionist focus to more holistic systems views.

Chapter 10 is the conclusion. The three primary objectives of Deep Information are profiled against recent and emerging policy trends. Chapter 10 begins with a brief examination of why NEPA and its impact statement emphasis has often represented the antithesis of Deep Information. A review is made of the policy transformation of the legacy information systems toward sustainable objectives and the gaps which still exist. Finally, the formal organizational dimensions of information policy are engaged, including the momentum for a National Institute of the Environment.

# REFERENCES

Berke, P., & Kartez, J. *Sustainable Development as a Guide to Community Land Use.* Chapel Hill, NC U.N.C. Center for Urban and Regional Studies, 1995.

Carson, R. (1962) *Silent Spring.* Boston: Houghton Mifflin.

Culhane, P., et al. *Forecasts and Environmental Decision Making: The Contents and Accuracy of Environmental Impact Statements.* Evanston, IL: Center for Urban Affairs and Policy Research, Northwestern University, 1985.

EPA Ecosystem Protection Workgroup.*Toward a Place Driven Approach: The Edgewater Consensus on an EPA Stategy for Ecosystem Protection.* Washington, D.C.: Environmental Protection Agency, 1995.

Karplus, W."The Spectrum of Mathematical Models." *Perspectives in Computing,* 3 (2) (1983): 4–13.

Lee, K.*Compass and Gyroscope: Integrating Science and Politics for the Environment.* Washington,D.C.: Island Press, 1993.

NJ Department of Environmental Protection. "GIS Partnerships Celebrated," *N.J. GIS Update,* 27, (Summer 1995): 1–2.

Simon, H. A. *Administrative Behavior.* 3rd ed. New York: The Free Press, 1976.

Westbrook, R. *John Dewey and American Democracy.* Ithaca, NY: Cornell University Press, 1990.

Yankelovich, D. *Coming to Public Judgment: Making Democracy Work in a Complex World.* Syracuse, NY: Syracuse University Press, 1991.

# 2

## Visualizing Environmental Information Policy

....................

### DIAGRAMMING OVERVIEW

*I*n Chapter 1, we were introduced to the highly complex dynamic interaction between human socio-economic systems and natural systems. A series of generalized diagrams were used to visualize basic systems and categories of exchanges. Since a primary objective of systems analysis is improved understanding of process dynamics, flow diagramming has evolved to provide a simple yet powerful analytical tool. Three major discipline-based diagramming perspectives have direct relevance to the arena of environmental information policy. These include environmental systems, information systems, and policy systems. As one would expect, each field has evolved a unique visual jargon of representational formats and conventions.

Environmental systems analysts investigate and describe a broad spectrum of phenomena, ranging from global climate to insect life cycles. Visualization has played an integral role throughout the history of science and technology. Figure 2.1 depicts the general hydrologic process water cycle. A version of this pictorial depiction was used as early as the 1400s. Figure 2.1 contains the fundamental systems concepts of cascading exchanges, feedback cycles, and parallelism. Present-day modelers often redraw this figure as a schematic mass-flow diagram with every physical component, such as surface vegetation, represented as a rectangle.

Environmental system modelers make mathematical predictions of environmental flows such as storm water runoff, acid rain deposition, and species' populations. Throughout the century, civil engineers have studied and designed systems such as metropolitan water supplies. Prior to the advent of digital computing, they relied heavily on field observations, laboratory experiments, and analog (electric circuit) models to develop simulations. Following World War II, ecologists began to adopt systems analysis approaches to develop models of biological systems. Due to the array of exchanges in a complex system, these models often focus on a critical exchange, such as energy or nutrients. Figure 2.2 is an example of an ecosystem modelling diagram. The use of a consistent symbol set provides a powerful communication device for symbol-literate specialists.

From its origins in electrical engineering circuit diagrams, to the logic flows depicted by programmers, to the tools of today's system professionals, information analysts have a rich heritage of flow diagramming. Data flow diagramming (DFD) is

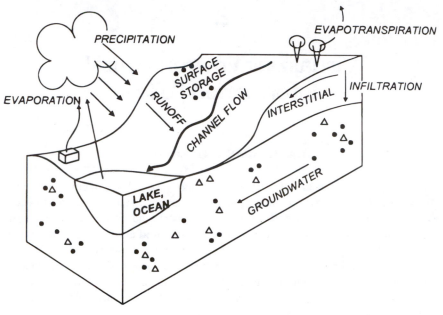

**Figure 2.1. Hydrologic Cycle**

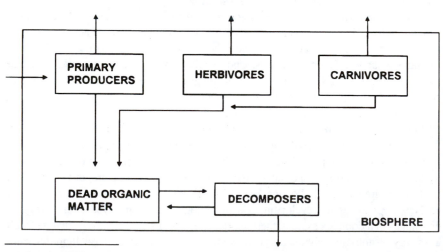

a based on White p. 77

**Figure 2.2. Ecological Model**[a]

a standard approach used in documenting current information paths and processes within and between organizations. Coupled with the development of performance objectives, DFD serves as the basis for the development of re-engineered flows.

Traditional bureaucratic organizations have often been established as hierarchal pyramids segmented both horizontally and vertically. Horizontal divisions represent functional specialization. For example, in a public environmental agency it is common to have divisions focusing on air, water, and hazardous waste. The vertical strata often are in three layers: transactional or production, tactical middle management, and strategic top management. While all levels rely on a combination of internal and external data sources, the importance of external data traditionally increased for tactical and strategic functions. Figure 2.3 is a representation of the differing information system needs at the various levels of a bureaucratic organization (Burch & Grudaitsui, 1986) Although advances in information systems are accelerating the evolution of formal organizations to more horizontal and flexible structures, many of the basic information needs remain.

Conceptual DFDs, as shown in Figure 2.4, provide the framework for identifying each stage and location of information generation, processing, use, and ultimate disposal. In this simplified example, the permit review activities are all carried out at the transactional level while the information systems are maintained at the tactical level. The organization is divided into office and field staffs. In this form of box and arrow diagram the figure's key depicts the basic components of the process: data sources and destinations, data flow, data process or transform, and data store. In order to implement an actual information system, the conceptual data flows need to be translated first into

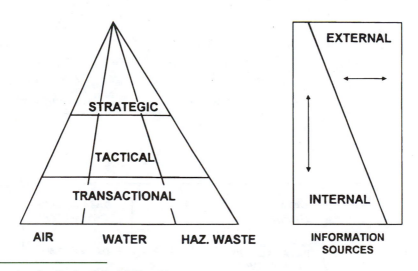

---

**Figure 2.3. Bureaucratic Information Hierarchy[a]**

logical data flows, and data must be organized in a format consistent with the system's logical program operators. For example the facility record management could be auto-mated by means of a relational database, while the map operations could be done with a Geographic Information System (GIS).

Although many have likened policy development to the making of sausage (some-

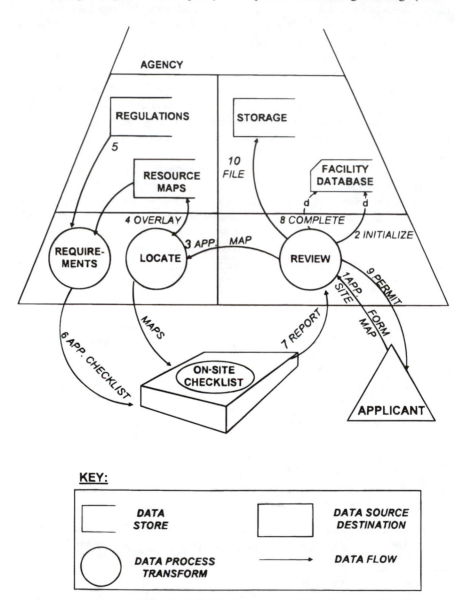

**Figure 2.4.  Data Flow Diagram For a Permit**

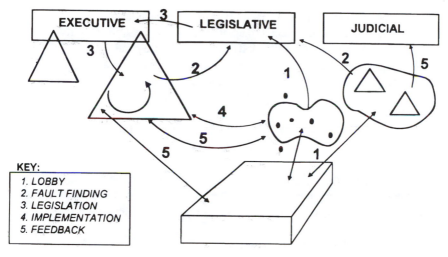

**Figure 2.5.  General Policy Process**

thing you don't want to see), the complex flows over time coupled with multiple decision points require simplification and organization to understand underlying patterns.
McClure has advocated that diagrams should play a key role in conducting information policy analysis (McClure, 1989). Figure 2.5 shows a macro-scale general flow of
a common policy development process. The basic components are the three branches
of government and multiple publics. An interest group may perceive a problem or
opportunity and advocate the need for a program (1). A legislative committee may conduct a policy analysis and obtain input from the executive branch and affected publics
(2). If a program is legislated and not vetoed it is assigned to an agency for implementation (3). The agency then promulgates internal procedures and external rules and regulations. The latter have a comment/review feedback cycle. When fully legitimized, the
program is implemented (4). Program evolution may have multiple pathways. The
implementing agency may monitor and make internal adjustments; the Executive
Branch or the legislature may audit and instigate changes; and an opposing interest
group may seek judicial relief (5).

Examining all of these selected examples, we see the commonality of a basic "box
and arrow" structure where the box represents a component which functions as a source,
transformer, or depository, and the arrow indicates an exchange between components.
We should note that this approach represents a useful but highly simplified mechanistic
modelling of what are typically quite organic functions. Another diagramming characteristic that reflects the underlying nature of most systems is that of scale. Each component of a system is usually composed of subsystems, such as the tree in Figure 2.1 , the
resource maps in Figure 2.4, and the legislature in Figure 2.5. It is important to be able
to "zoom" in and out to get macro, local, and micro views of systems, subsystems, and
subsubsystems. The scales are not just bigger or smaller clones. At each scale, a differ-

ent phenomen or dominant set of characteristics may be significant.

Finally, in selecting a system to diagram we must necessarily simplify by isolating it from its contextual universe. Good analytical practice requires the articulation of such simplifying assumptions. The representation of these connections to externalities is commonly shown as bounds or the surrounding universe.

## THE BASIC ENVIRONMENTAL INFORMATION FLOW DIAGRAM

The basic diagram that will be used in the rest of this monograph to describe and critique information policy integrates the elements from Chapter 1 and the three major visualization sources-environmental systems, information systems, and policy analysis-discussed above. The diagram is shown in Figure 2.6.

It has two major sectors, socio-economic systems and the physical environment. The latter contains an immense web of subsystem cycles that are represented by the pictorial icons of natural and human activity subsystems. The environment is shown as a pictorial bird's eye cutaway (block) diagram. In contrast to the abstractness of Figure 2.1, this approach illustrates both the critical importance of spatial relationships (adjacency, proximity, distance), and, as Karplus indicated, the vast differences in available data between surface and underground attributes. The indication of a timeline serves as a reminder that the environment is constantly in flux. This is important to both our understanding of dynamics, and the varying rates of information obsolescence.

The socio-economic sector is divided into governmental and private subsectors. The former includes the three constitutional branches as well as the bureaucratic agencies of the Executive, which some political scientists have classified as a "fourth branch." The United States and many other nations are a federalist system of government. State and local governments constitute a rich hierarchy of rights, responsibilities, and associated information flows.

The private sector consists of an overlapping set of citizens and organizations. The former are primarily defined by spatial jurisdictional boundaries, while the latter are the cumulative result of both legal and individual actions. It is useful to differentiate between for-profit corporations, such as a manufacturer, and non-profit, non-governmental organizations (NGOs), such as an environmental interest group. In addition to formal organizations, there are numerous informal collections of people with similar interests, ranging from neighbors to Internet list groups.

The interface between the socio-economic systems and the physical environment consists of two primary overlays. All of the land is owned by citizens, corporations, or a unit of government. The owners have primary management control of site development and management of human activities, as well as the stewardship or destruction of natural resources. All of the land is located in one or more governmental jurisdictions. These spatial delineations determine the distribution of programs, both service delivery and regulatory.

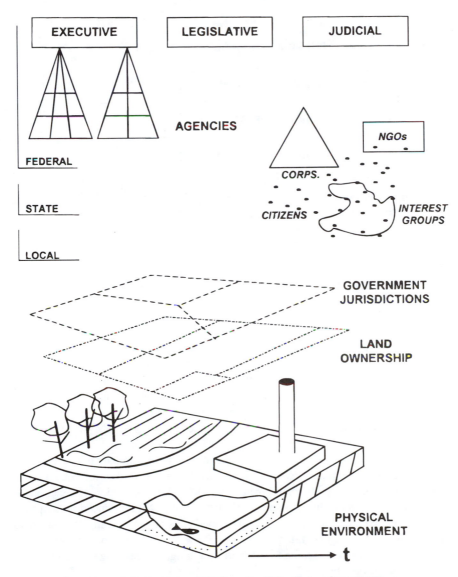

**Figure 2.6. Environmental information Policy System Components**

The above descriptions cover the "boxes" of the system diagram. As introduced in Chapter 1, there are two basic sets of flows. Physical exchanges to the environment will be represented by heavy arrows. These can either be from a landowner, as in a construction or emission activity, or within the environment, such as erosion deposition in a lake. Information flows can be between socio-economic components, such as the review of a draft environmental impact statement, or between a socio-economic com-

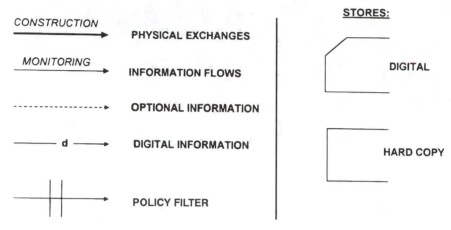

Figure 2.7.  Flow and Store Symbols

ponent and the physical environment, such as monitoring stack emissions. Flows are labelled to denote character.

Figure 2.7 shows some of the primary types of information policy-related flow elements. Heavy arrows are physical flows while light arrows are informational. A dashed arrow represents an optional flow such as a comment on an impact statement (formal option), or an unconducted mandatory site inspection (informal option). A major focus of information policy is in the continued conversion from hard copy to digital formats. Information arrows with a "d" represent the latter. Within organizations, information storage and processing can be either manual (hard copy) or computer assisted. Some of the primary information resource management subsystems are databases (relational, full text, etc.), and GISs. These are shown as symbols.

Information policies affect generation, processing, dissemination, and disposal. From a flow diagramming perspective these may depicted as pumps and conduits, and filters. Pumps and conduits control the content, format, and direction of the information flow. For example, the Community Right To Know Act introduced in Chapter 1 requires that commercial activities of a specified class annually complete a standard form stating the amount and storage location of chemicals classified by a formal coding system. This report is to be transmitted to a designated state agency, and a local emergency response board.

Information policy filters limit the content and/or distribution of the flow. Formal filters specify constraints. FOIA has nine "exclusions" ranging from "geologic information" to "national security" by which citizen requests for national agency information will be denied. Informal filters include access and resource constraints, such as an interest group not knowing about an imminent decision, a small business not being able to understand reporting requirements, or a citizen not having ready access to the Internet. Additional informal policy filters arise from agency values. An example is the internal decision whether or not to monitor conformance to a construction permit.

Although models by definition are a simplification, inherent in analyzing actual systems is a tendency for complexity. Combining Figures 2.6 and 2.7 can quickly generate a visual gridlock of spaghetti. Three approaches will be used to manage this problem. First, only those socio-economic components directly relevant to the issue being discussed will be depicted. Second, most processes are sequential. Therefore, the flows will be numbered. If space is available, the flows will be labelled on the arrow. If not, a legend will be used. Finally, many processes are extended over time. For example, the legitimization of a policy process precedes its implementation, which (sometimes) is followed by a review. The underlying three-stage sequence can be depicted on a timeline diagram. The detailed flows can then best be illustrated as a time series of "snapshots."

Actors are considered to be organizations and individuals who formally participate in decision-making. For public decisions they have legitimized standing. For private decisions they either have resource ownership, such as land, or social pressure influence. Stakeholders are those organizations and individuals that are directly affected by these decisions. In this example (and throughout the volume), environmental subsystems such a habitats will also be considered stakeholders. This is a central element of "Deep Information."

The diagram has three major zones: government, publics, and the physical environment. Government has the three basic branches: judicial, legislative, and administrative. The latter is expanded to show the agency bureaucracies (the fourth branch). Bureaucracies are depicted as hierarchial triangles layered from the top down as: strategic, tactical, and transactional (Burch & Grudnitski, 1986), and vertical separations representing topical specialization. To represent the shared powers in our Federal system, the basic governmental construct is replicated at national, state, county, and local levels.

The publics are a heterogeneous group of all non-governmental human actors and stakeholders. These include individual citizens, private firms and corporations, formal interest groups, ad hoc interest groups, and the media. The physical environment is depicted as a block diagram (a tilted three-dimensional cutout). This permits a view of surface activity and a partial view of subsurface phenomena. The fourth dimension of flows and change states can be represented either by arrows or by a time series of views, as in frames of a movie film.

Information flows are shown as arrows connecting two or more parts of the diagram. Two basic types of arrow are used: a heavy arrow for physical changes to the environment, and a light arrow for information exchanges. The latter include public governmental records, private exchanges, and information published or broadcast by the media. Labels on an arrow summarize the content.

## INFORMATION POLICY

Information flows constitute a major determinant of our individual and collective environmental understanding. The flows are not ad hoc. In our highly organized society, information policies directly affect the quality and outcome of most environmental decisions. In Chapter 5, we will look in depth at three arenas of environmental information policy. Generation policy deals with the creation of data which describes the physical environment, human activities which affect the environment, and the subsequent analysis of this data to develop information. Distribution policy addresses access to and dissemination of environmental information. Stewardship policy focuses on the quality and currency of the information, its increasingly collaborative construction, and ultimate archiving or disposal.

These policy arenas are clearly interrelated. Historically, a majority of environmental information was generated for internal consumption by a public bureaucratic program. With the advent of the modern environmental movement, a rapid policy evolution took place toward open access to program information. NEPA calls for the active interagency and public review of significant environmental projects prior to final decision making. With the recent advent of networked information systems, the increasing reliance on dynamically shared data through such constructs as a clearinghouse, and the geometric explosion in the amount of environmental information, stewardship policies reflecting comprehensive approaches to information resource management (IRM) have begun to emerge.

Since virtually all public programs involve information, there are a myriad of existing information-related policies. Public information policy creation is primarily driven by two forces: special interest groups seeking to change a particular policy, and administrators and legislators attempting to make information handling more efficient and cost-effective. Examples of the former include proposals by some groups to increase chemical release reporting by companies, and by other groups to reduce regulatory reporting burdens. Examples of the latter include the promulgation of Federal Information Processing Standards (FIPS), such as the standard industrial classification system (SIC, Office of Management and Budget, 1987). These pushes and pulls are highly incremental.

The emergence of policy approaches to information stewardship as a response to the ongoing electronic information revolution has the potential to facilitate environmental sustainability. The realization of this symbiotic relationship, however, is in no way ensured. Over the centuries we have developed a number of distinct, institutionalized, high-inertia, legacy information subsystems that reflect varying facets of then-prevailing environmental viewpoints. Collectively, these subsystems constitute major pragmatic and philosophical barriers to sustainability. The incremental nature of most information policy actions often serves to perpetuate these barriers. In the next section, we will utilize the basic dataflow constructs to examine a hypothetical scenario that illustrates these historical subsystems and barriers to knowledge building and collective action.

## REFERENCES

Burch, J.,& Gary Grudnitski. *Information Systems*. New York: Wiley, 1986.
McClure, C., "Frameworks for Studying Federal Information Policies: The Role of Graphic

Modelling," in *United States Government Information Policies: Views and Perspectives*, edited by C. McClure, P. Hernon, H. Relyea. Norwood, NJ: Ablex, 1989.

Office of Management and Budget. *Standard Industrial Classification Manual*. Washington, D.C.: NTIS, 1987.

White, I., D. Mottershead, and S. Harrison. *Environmental Systems*. New York: Chapman Hall, 1992.

# 3

........................................................................................................................................

# Greenetowne—
# An Information-Based
# Environmental History

........................

## INTRODUCTION

*I* n Chapter 1, the proposition was advanced that many of our current activities
are not environmentally sustainable. These behaviors are perpetuated in part
due to our collective inabilities as actors and stakeholders to both develop a
deep understanding of the environment and utilize the checks and balances of the
public sector, the market forces of the private sector, and the ethical behavior of
individuals to reestablish a more robust equilibrium. Information policy constitutes
a major arena of both barriers to, and opportunities for, the transition to sustainability. To "ground" these somewhat abstract thoughts, it is useful to examine a
hypothetical historical scenario.

Although many associate sustainability with world-scale issues like global warming, it is most readily engaged locally. Greenetowne is a community in America's heart-
land. Like all communities, its health and prosperity is in part a function of sustaining
the quantity and quality of critical environmental subsystems. For Greenetowne, two
such subsystems are its drinking water supply and its wetland habitats.

We are going to "look" at Greenetowne over its hundred-plus years of occupa-
tion and development following the displacement of Native Americans. History will
be viewed as a series of connected episodic periods, often dominated by major trig-
gering events.

## PHASE I: SETTLEMENT

White traders and trappers travelled through this region of rolling hills and upland
forests, beginning in the 1820s. They made transport use of the Blue River and occa-
sionally stopped at the small seasonal Native American settlement located on a ter-
race overlooking a stretch of river rapids. The original inhabitants had fished the river

and tributaries, hunted the marshes and uplands, gathered wood from the forest, and farmed the floodplains and terraces for hundreds of years. Their population had grown and shrunk based on natural conditions and interactions with other communities. A Federal geological survey team used the Blue River on its way west in 1832, but made no mention of potential local mineral resources. Some white settlers had already moved into the region while Federal troops were removing the Native Americans in the late 1830s.

In 1840, the state legislature incorporated Browne County, adopting a survey which included the delineation of the county bounds, and the creation of a county seat, Greenetowne, on the north side of the Blue River at the rapids. The legislation also established a county court to be located at Greenetowne to facilitate the local conduct of state business. The basic size of the county was determined by the radius of a daytrip on horseback. A charter for the development of a county legislature was passed.

In order to encourage settlers and provide an orderly transfer of land ownership from the state to private interests, a subdivision survey map (plat) was created. Utilizing the "modern" Jeffersonian approach of systematic geometry, it ignored the area's landforms features (except the river) and created six essentially square towns. Each town was broken into four sections, and each section had four great blocks. Towns were named. Sections and blocks were uniquely identified in a standard numerical order. One subsection of a central block in each town was reserved for public purposes, such as a school. The plat was filed at the court, and a land auction was held. For each sale, a deed was recorded at the court. Some purchasers of blocks were banks and speculators who subsequently subdivided them into farm and building lots. In each case, the plats and deeds were filed. Each deed referred to the platted town–section–block–(subdivision) survey.

Frequently, a purchase was made with borrowed money. The lender was promised rights in the property as collateral to default on the loan. These mortgages (liens) were also publicly filed at the court. The land records served as the basis for real property taxation, the primary source of county and town revenue. Assessment was based on both the value of the land transferred in the deed and the improvements, such as buildings. Figure 3.1 shows a composite map of the initial creation of jurisdictions and ownership parcels. Figure 3.2 depicts the dataflows associated with a land transaction. The set of land ownership record systems, including deeds, mortgages, plats, and assessments, constitutes the "cadastre." The cadastre was developed and maintained solely at the county level.

Settlers all used water and generated waterborne wastes. Water was drawn directly from the river at multiple points. Away from the river, creeks, shallow dug wells, and rain barrels were used. Human wastes were discharged into pit privies. Used process water and factory waste from the river-edge lumber and flour mills, and a tannery, were piped directly back to the river. Within twenty years, most of the forest cover had been cleared for agriculture, fuel, and construction. Surface runoff carried eroded soil and domestic animal wastes into channels and roadside gutters. Sediments were deposited in upland streams, ponds, and wetlands. Much of this material ultimately ended in the Blue River.

Figure 3.3 depicts the water-related flows of this initial settlement period. Browne

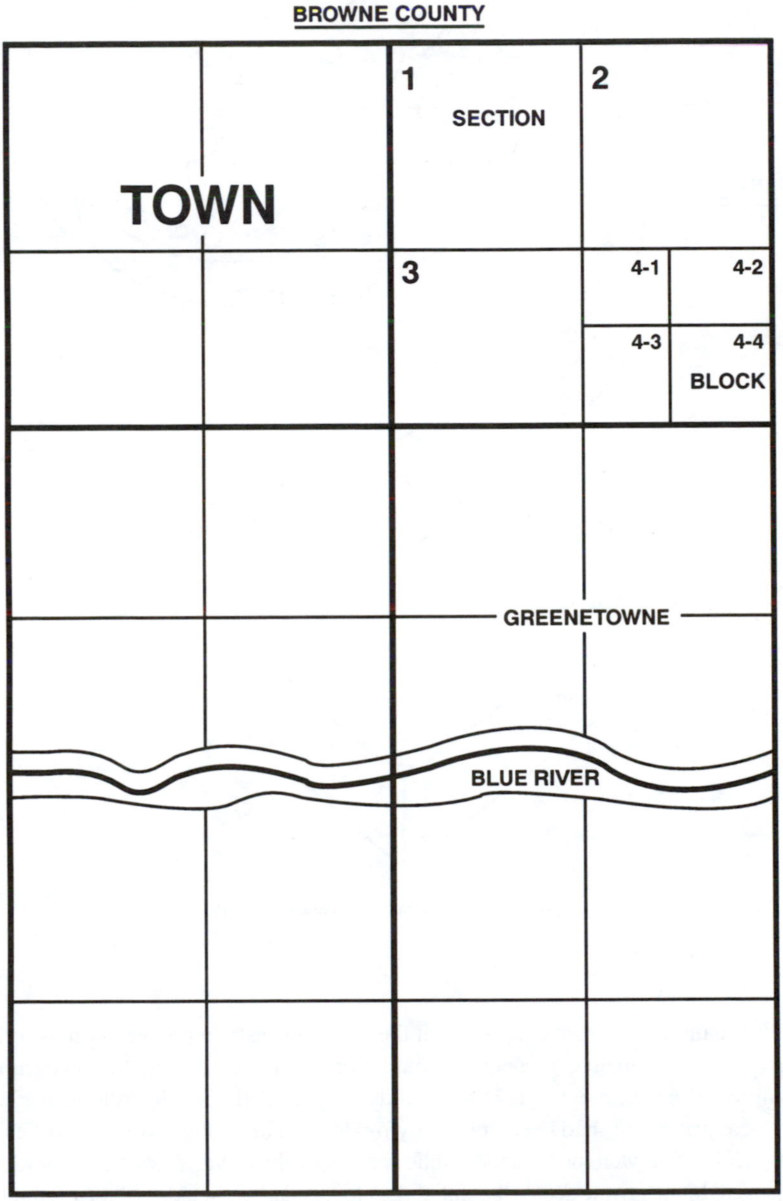

**BROWNE COUNTY**

Figure 3.1. Local Jurisdiction and Initial Ownership Parcels

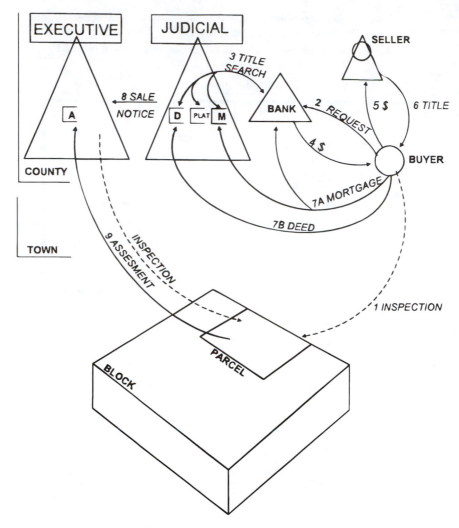

**Figure 3.2.  Land Sale Information Flows**

County's natural resources coupled with the public cadastre were the springboard for private sector commerce to generate major physical changes in the environment. Within two decades, two thirds of the county was "settled." The hydrologic cycles of most local watersheds had been drastically modified. The changes were almost entirely physical. There were no specific public records of these water, waste, and sediment exchanges. In the private sector the initial developers had knowledge of the original on-site water and waste facility construction. Whether or not this information was retained

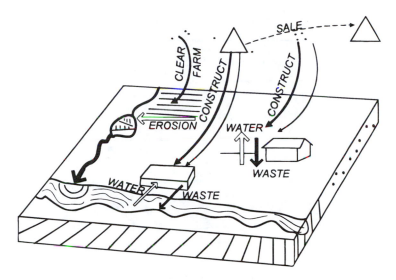

**Figure 3.3. Initial Water Related Flows**

and transmitted to subsequent owners was purely ad hoc.

Throughout this period of rapid settlement, the community's value perspective regarding the environment was characterized by a set of complementary views. First, it was seen as a commodity, the major resource in a primary economy. Second, some features were viewed as problems to be overcome by "civilized" people. These included the dense forest, swamps, and wolves. Finally, some phenomena, such as tornadoes, floods, and lightning fires, were lumped as "acts of God."

A distinct duality emerged regarding the common knowledge of the environment. On one hand, the geometrically drafted plats and their associated documents constituted the official "information reality" describing the physical world. In contrast was the "physical reality" of cleared forests, eroding farmlands, and hundreds of wells, privies, and discharge pipes. Unlike the farms and buildings which were readily visible to all, and quickly became the collective image of the community, the erosion and sedimentation was highly dispersed and typically occurred in remote locations. The privies and sewer discharges, except for a few riverside operations, were completely buried out of sight.

## PHASE II: CHOLERA OUTBREAK

Browne County suffered a local economic depression preceding and during the Civil War. In the post-war period, development pressure grew rapidly. The decennial Federal census of population tracked each shift. In 1870, the residents and businesses in Section 3 Blocks 1 and 3 of Greenetowne successfully petitioned the state legislature

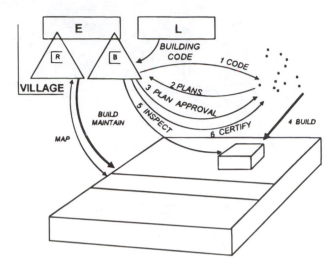

**Figure 3.4. Village Development**

to create a village. It was named Rapids. The village created a Roads Department which used increased property taxes to improve roadways and roadside drainage. A number of small riparian wetlands were filled to increase riverside building lots. A simple building code was passed. It controlled the design and separation of new structures to reduce the threat of fires which had plagued the community. The road superintendent kept a file of road-related maps, while the building inspector established a file of approved building plans. The number of residences and businesses in the village grew rapidly. Figure 3.4 shows the building and road activities.

In 1885, a cholera epidemic broke out in Rapids and parts of Greenetowne. Due in large part to the energy of the local newspaper, within weeks the State Health Department established a Cholera Commission to investigate. The Commission utilized scientific knowledge and methods which had only recently become widely available. In the late 1860s, scientists and doctors in London confirmed the germ theory of disease by definitively linking a cholera outbreak with a public well polluted by human wastes. In Rapids and Greenetowne, the investigators found a number of polluted dug wells and clear indications of human wastes in the vicinity of some riverfront water intakes.

Utilizing its broad police powers, the Commission ordered the construction of a centralized water supply. It was to take water from the river upstream of development and divert it to a holding reservoir. From here it was to be piped to the village and adjacent town. A County Water Utility District, a special-purpose unit of government, was created by the state legislature to build and operate the system. The District was funded by revenue bonds to be paid off by a property tax and users' fees. It kept maps of its system and records of billings.

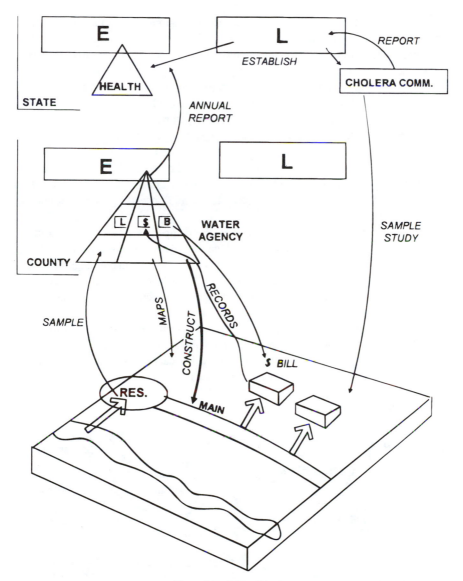

**Figure 3.5.  Water District**

In addition, the District was required by the State to set up a laboratory and daily test the water supply. Tests included color and smell as well as chemistry and biology. It was impractical to test directly for pathogens, so a proxy indicator species was specified. E-Coli, a bacterium which exists in great numbers in our digestive tracts, was easy to culture. Its presence represented the probability of human waste, and thus the probability of waterborne disease. Test records were summarized in an annual report to the state. The major flows related to these actions are depicted in Figure 3.5.

At the urging of the business community, and with the strong support of the news-paper, the new water system was also designed to incorporate the flows and pressures necessary to adequately handle fire fighting demands. This was significant not only for public safety but also in terms of fire insurance rates, which had been increasing rapid-ly. In 1895, a private insurance mapping company, utilized collectively by underwrit-ers, conducted a detailed analysis of the village. The process had two stages. First, village and county sources were reviewed to create a current basemap of roads, parcels, owners, and building construction. This was followed by detailed site investigations to update the public information and enumerate additional data regarding site activities, such as manufacturing processes and use and storage of incendiary materials, and fire safety equipment.

Although voluntary, all owners allowed the company's staff on-site and inside buildings, because unless they were mapped they could not obtain fire insurance. Some initial holdouts quickly agreed due to informal peer pressure (both buildings and neigh-borhoods were rated), and newly adopted formal policies of the local banks that required fire insurance as a condition of mortgages and business loans. The new water system qualified properties in the village for significant savings.

The fire insurance maps were primarily sold to insurance companies but were also available to the public for sale. The new village library, that was built and operated as a private non-profit organization, purchased a set. The data flows related to this process are shown in Figure 3.6. This private sector process was purely informational.

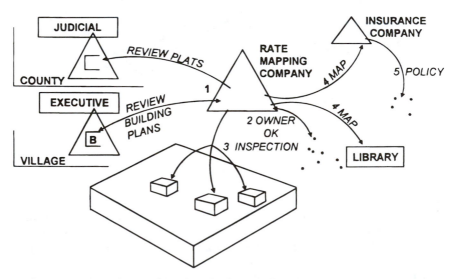

**Figure 3.6. Fire Insurance Ratings**

From the perspective of community values, this period marks the reawakening of consciousness regarding the critical relationships between natural and human systems—an awareness that was a daily fact for early homesteaders but had quickly been lost in the village. This reawakening was actualized as a form of logical positivism. The response had three facets. First was the utilization of the new tools of technology and science to control nature through the development of a public works infrastructure. Second was the expansion of the regulatory powers of the government to limit the range of human activities on private land to protect and improve the public health, safety, and general welfare. Finally, the perception of problems and the framing of their solutions was reductionist, a form of "divide and conquer".

The amount and scope of environmentally-related information generated in this period represented an exponential increase from the settlement period. The paper records and maps were created and kept by a growing set of public and private organizations. A composite of Figures 3.4, 3.5, and 3.6 would not be readily comprehended.

## PHASE III: INCREMENTAL GROWTH AND A GREAT STENCH

Over the first half of the 20th Century, the town and county exhibited punctuated periods of growth and change, reflecting larger trends in the region and nation. A rail spur was constructed across the northwestern corner of the county. This in turn generated a cluster of agricultural and manufacturing activities in an unincorporated area, which became locally known as "Sheds." During World War I, a railside munitions factory was built and operated. Based on active farmer education by the state agricultural college, hundreds of acres of wetlands and low-lying soils were converted to agriculture through extensive construction of ditches and the installation of networks of drain tiles.

During the early years of the 1920s, a protracted drought severely reduced flows in the Blue River. Many of the shallow wells went dry. In August of 1923, massive algal blooms in the river's pools signalled its transition to a deoxygenated eutrophic state. The newspaper decried the "Great Stench" and those that could temporarily left the village or sent their children to stay with relatives. A major health alert was promulgated.

Response to the crisis was twofold. The State Health Department outlawed the discharge of raw sewage to the river. It mandated the construction of an interceptor sewer, and a sewage treatment plant below the rapids downstream of the village. The plant was to provide "primary treatment," the removal of settleable solids, followed by chlorination. The solids were to be dried and disposed of in a dump. With state legislative approval, the county created a sewer district to implement and manage the facility.

In a parallel action, the county hired a consulting geologist to develop a plan for an alternative water supply. The firm consulted a general bedrock geology map prepared by a doctoral student at the state university and a U.S. Geological Survey (USGS) map which depicted contours and landforms. They drove around the county examining rock outcrops and streams, and examined numerous existing wells. For most of the latter, the only available historical data was anecdotal. The consultants

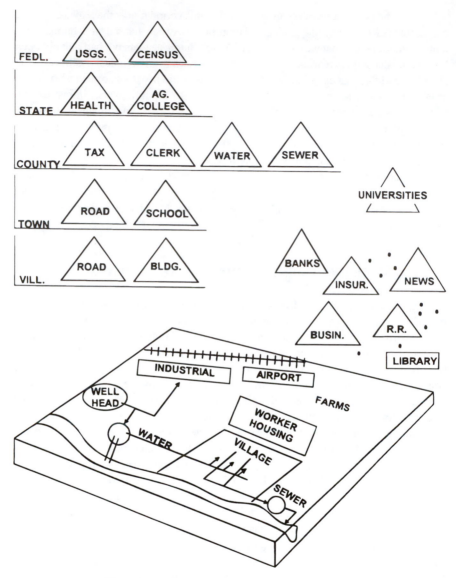

**Figure 3.7.   WWII-Vintage County Information Holders**

hypothesized that a significant aquifer existed in the central western region of the county. Funds for a test well were authorized. It was successful. The Water Utility District was then authorized to issue revenue bonds for the construction of a group of deep wells and a transmission-distribution system. The entire county was included in the expanded District.

The Depression was a period of major decline. Businesses and farms closed. Population migrated away. Federal monies were used for a variety of projects. Channelization (clearing of vegetation, straightening, and lining) and levee construction were undertaken for numerous floodprone streams. State and Federal agencies coordinated the creation of an agricultural soil survey and map as part of a farmer outreach program to reduce erosion and improve crop yields. The USGS established a permanent gauging station on Blue River as part of a nationwide data collection network.

During World War II, the industrial complex on the rail siding was greatly expanded. Nearby, a small airport was constructed. A major complex of temporary housing for workers was developed just north of the village.

Along with the physical change and urbanization, community environmental values evolved in this period from a shared collective view to a plural collage. Many of the original families were gone. New residents came from other regions with a diverse mix of rural and urban backgrounds. The history of the founding, the abundance of fish and wildlife, the cholera, and even the more recent stench was not shared. A majority took it for granted that you got water from the faucet and human wastes went down the toilet, end of subject.

From an environmental information perspective, the earlier trend of isolated exponential increase continued. Each of the major actors managed a use-specific set of paper records of varying vintages. Many stakeholders had little or no knowledge of or access to these information sets. Figure 3.7 shows a view of the county and the primary information holders. Flow arrows are not attempted.

## PHASE IV: SUBURBAN BOOM AND
## ENVIRONMENTAL AWAKENING

Postwar prosperity brought rapid growth to the region. Driven by pent-up demand and low cost veterans' loans, major clusters of residential subdivisions began spurting up in the agricultural areas. These developments had quarter and half acre lots, public water, and on-site septic systems. Costs were minimized through the use of standardized mass construction methods. Concern rose among long-term residents about the loss of community "character." This concern quickly focussed on public health, as many of the septic systems "failed" during spring high water tables, leaving seasonal pools of sewage in many backyards.

Problems with on-site sewage disposal were endemic across the state. Legislation was passed amending the Health Law to create a regulatory septic system permit process. It included specific standards for location in relation to structures, surface water, and wells. Soil tests were mandated to estimate location of bedrock and groundwater, as well as permeability. Design and construction standards were set. The new County Health Department was delegated the responsibility to manage the permit system. Data flows for these permits follow the generic process shown in Figure 2.4. Three major results of this action were a significant increase in suburban lot sizes to a one acre minimum, a shift of development to areas with well-drained sandy soils, and pressure from landowners in areas of poorly-drained soils to extend the sewer system.

The response to initial sprawl transcended health concerns. By this time, each of the originally platted townships had incorporated as towns. Each town in the county, with Federal aid, developed a municipal Master Plan and associated zoning map and ordinance. Landuses were spatially segregated among commercial, industrial, residential of varying density, public, and agricultural activities. One basic information set used by the planners was a new Federal soil survey of the county. This map and report addressed both agricultural and suburban development soil interpretations.

The conversion of fields to impervious suburban roofs and paving increased the historical flooding problem. The new plans and ordinances addressed this by requiring structural solutions in the form of systems of storm drain pipes and culverts. Little was left of the original hydrologic system of streams, wetlands, natural floodplains, and associated locally high groundwater tables.

Although the local plans called for future concentrated development to be focused around Rapids, this was superseded when the State Department of Transportation located an Interstate highway parallel to the railroad in the northern sector of the county. Local political and business pressure subsequently gained authorization for an interchange near Sheds. Within a few years, the region's first shopping center was under construction there. The pro-development interests were also looking at multi-modal transportation. After numerous feasibility studies and county initiatives, the state created a County Airport Authority. This new body issued revenue bonds and began the total rehabilitation of the private air field into a regional air terminal.

Despite suburbanization, agriculture was alive and well in the County. Small farms merged into a few large operations. Major introductions of fertilizers and herbicides coupled with increased mechanization increased productivity. Free technical assistance, such as the new soil survey, and a series of Federal crop support programs were major contributors to revitalization of this sector.

The rapid development spawned many critics who were concerned about the radical changes in landscapes, vegetation, wildlife, streams, and wetlands, and the rise in apparent air and noise pollution. Some individuals gained membership in planning and zoning boards to pursue their agendas from within the "system." Another segment formed a local chapter of a national environmental organization. Still others created a local "land trust" whose primary objective was to protect sensitive areas by acquiring ownership title or easements to the property.

In the late 1960s, spurred by new Federal and state water quality legislation, a series of chemical analyses were conducted for the Blue River. The tests revealed extremely high levels of organic and inorganic chemicals. The sewage plant was mandated to upgrade its level of treatment to "secondary," entailing the addition of biological treatment that was to remove approximately 80% of the effluent's oxygen demand. Some riverfront industries were required to provide specialized pretreatment prior to discharging into the municipal sewer main. Two facilities closed operations and moved to another state. Upon further investigation of the river's bottom sediments the state determined that, due to long-term buildup of heavy metals, the river could no longer be used as a domestic water supply.

Community environmental values in this period began to polarize. At first there was overwhelming support for planned growth. This was followed by a series of highly publicized "events" in which small groups of environmentalists opposed major new developments. These events were brought directly into area homes "live" by Rapids' new television station. The newspaper had a continuing series on both the issues and the confrontations. By the end of the period, a news poll showed that the vast majority of residents felt that environmental quality was important and that stronger action was needed to protect it.

From an environmental information perspective, the disconnected single-purpose scientific or managerial data sets of the previous period continued to proliferate. Although there was much new environmentally-related information there was little or no environmental knowledge generated of how systems functioned. In contrast, the state had mounted a transportation study of land use and traffic flow in order to plan the Interstate. The study made major use of a new technology, computers, to handle massive amounts of data and computations. This model included serious faults, particularly regarding its self-fulfilling future growth forecasts and its total disregard of environmental value in its benefit-cost submodel. However, it was based on considerable process research and was a bellwether of a comprehensive approach to managing a complex system.

## PHASE V: LATE-MODERN "SUPER" PROBLEMS

Starting with the NEPA, the 1970s and 1980s were a period of rapid proliferation of Federal and state legislation regarding the environment. Air, water, hazardous waste, endangered species, and wetlands were but some of the topics covered by "command and control" regulations which dictated pollution control solutions. Some laws were implemented directly by the Federal government, while others became the responsibility of the state.

In Browne County, a routine 1983 water quality sample for the airport shopping center revealed the presence of newly regulated carcinogens. State and Federal authorities were called in and began a detailed site audit including a set of monitoring wells. It was soon established that the probable source of the pollutant was the old munitions facility and associated businesses. The local well was condemned, and a connection to the county water supply was hastily constructed. Within a year, the area was declared a "superfund" national priority hazardous waste site.

Each stage of this crisis was carried in detail in the press and on TV. Remediation of the superfund site dragged out for a decade. Litigation attempted to unravel liability among all those parties, including lending institutions, who had been in the chains of title. Meanwhile, technical experts from different sides debated the probable distribution and path of the underground pollution plume using different computer models.

Environmentalists, alarmed at the health risk, began to question the safety of the county water supply. The State Geologic Survey was asked to delineate the approximate bounds of the major county aquifer. Based on a resolution of the county legislature and the state map, the Federal government subsequently approved the designation of a Browne County "Sole Source Aquifer." The designation brought with it some Federal resources for the further development of a groundwater flow model. This model revealed that in addition to pollution problems there may be long-term supply problems. Due to historical modifications of the landscape and streams, coupled with recent increased pumping, the consumption rate now far exceeds natural recharge.

Other Federal aid was used for dissemination of information to residents and businesses in the aquifer zone. In general, however, the Federal law had few teeth except stricter review of proposed new Federally-supported projects. The continued sustainability of the aquifer was in no way ensured.

Slowly, this situation engendered a new state of deeper environmental awareness and an emerging new paradigm of responsible stewardship. Each time in the community's history that its ignorance had damaged its water supply, there was an available "fix" through either spatial expansion, new technology, or regulatory and resource flow from higher levels of government. As part of the aquifer planning the county conducted an inventory of potential existing and future pollution sources. These included underground fuel tanks, septic systems, industrial chemicals, broadcast agricultural chemicals, and interstate highway, rail, and air transportation accidents. A preliminary map revealed that two thirds of the aquifer area and one half of its properties were now part of the problem.

The newscasts and public awareness campaigns about the aquifer were somewhat useful in maintaining a general awareness, but this waned as the superfund crisis receded in time. Although sensitivity levels were heightened during and immediately after the crisis, the underlying values of the community were still deeply divided. One faction felt that the system was at its breaking point and called for strong limits to new growth, elimination of many current practices, and extensive rehabilitation of existing facilities. Others felt the environment was actually quite robust and concluded that limited project review and remediation, such as the sole source aquifer program, were adequate. A third group (the largest) expressed concern about the environment but had no particular position regarding stewardship. All groups linked their environmental values to other values regarding the economy and lifestyle.

Regardless of the political debate about local courses of action, a major legal sea change had occurred. Pollution damage was no longer an externality or the sole domain of government. It was now the fiscal responsibility of landowners and process operators. As part of the post-NEPA legislation, citizens and NGOs had obtained legal standing and could bring suits against polluters, developers, and, increasingly, government agencies.

In contrast to the episodic crises of drinking water, the loss of wetlands over 100 years was a "silent crisis." Pressured by environmental groups who were alarmed over the accelerated loss of habitat and associated species, in the late 1970s the state passed a wetlands law that required permits for dredging or filling. Prior to implementation, the regulations called for approval of a wetland map. For Browne County, students and staff from the state university joined with local environmental organizations and public employees to complete the extensive field work.

In the 1980s the Federal government also enacted wetland controls. These had different definitions, maps, and permits. Both state and Federal approaches viewed wetlands on a one-time, site-by-site basis. The rate of loss was decreased but the net losses continued. The state amended its policy to "no net loss." For each acre filled, an acre of wetlands was to be created. The amelioration projects became uncontrolled, unmonitored experiments. Environmental groups bewailed the disjointed incrementalism of the isolated permit system decisions, and called for "cumulative impact analysis." This goal was never substantively addressed.

If drinking water and wetlands were not enough, a third pervasive environmental problem emerged in the 1980s: radon. Much of the county is underlain with limestone which contains traces of radioactive materials whose natural decay products include radon, a gas we cannot see, smell, or taste. If it can move through the overlying soil, and enters and is held in buildings, breathing this gas exposes the lungs to radioactive decay particles which can lead to lung cancer.

With Federal technical and financial aid, the State Health Department combined existing geologic mapping with a sample of voluntary confidential home testing to conclude that there was a high potential radon risk in portions of the county. This finding generated great alarm among citizens and in the real estate community. The radon phenomenon and subsequent alarm also occurred in many other areas throughout the state. In response, the state legislature passed a regulation which had two components. All public schools had to be tested and, if levels were above the Federal guidelines, remediated by sealing basements and installing ventilation systems. For homeowners, the law mandated the private testing and disclosure of results to the prospective purchaser prior to sale. No public record of this information process was developed.

## DEEP INFORMATION PERSPECTIVES

Each of the episodic drinking water crises, the continuing silent wetland crisis, and the newly revealed radon presence in Greenetowne represents functional ignorance of a complex environmental system, coupled with inappropriate private and/or governmental activity decisions. The current aquifer protection crisis can be seen as an information and knowledge crisis regarding the natural system, the social systems, and their linkages.

Deep information is founded on the shared development of knowledge regarding the functional processes of hierarchial environmental systems. Through most of Greenetowne's history the primary thrust of information generation was to support development, not to build systems knowledge. In recent decades, some systems analysis has begun. The crude computer models which began in the interstate highway period have continued to evolve. Other models are being developed and applied to surface runoff, groundwater, air pollution, traffic noise, and habitat evaluation. Such approaches are promising but limited due to the both information and process knowledge constraints. Regulatory and private sector decisions based on impact statements and modelling predictions therefore entail a large degree of error and uncertainty.

The second component of deep information entails the open, plural generation, analysis, and dissemination of environmental information. Measured against this objective, the history and institutions of Greenetowne are major failures. Numerous, narrow data gathering activities, coupled with closed internal manual information systems have resulted in a virtual environmental Tower of Babel. In addition, Browne County's socio-economic data does not constitute an integrated information system, having evolved little from its deed/mortgage origins. The explosion of local, state, and Federal environmental regulations has reinforced the historically generated disparate data islands.

The third component of "deep information" is its emphasis on downstream end-users, particularly local governments, and landowners and managers. As the recent environmental crises increased, the financial resource of the Federal and state governments declined. Local governments and property owners became aware that they were to play a central partnership role in pollution cleanup and the protection of important environmental resources. These actors have been poorly informed about the functioning of environmental systems, their stewardship responsibilities, and alternative courses of action.

The future of Greenetowne and Browne County is in serious doubt. A deep information approach for the region would entail the integration of the three traditionally isolated primary information sets: cadastre; regulatory; and scientific. These will be discussed in more detail in Chapters 6–9. In the next chapter we turn our attention to the first component of "deep information," systems knowledge building.

# 4

## Information, Knowledge, and Models

### ACTORS AND DECISIONS

*I*n the Greenetowne scenario, industry, government, interest groups, and individuals all made environmental decisions. From a perspective of rationalism, such decisions should be based on current information regarding the specific context, an understanding of the expected environmental outcomes, and an explicit awareness of social responsibilities. As we saw in the cases of drinking water, wetlands, and radon, the ideal of rationalism is not readily achieved. Part of the difficulty lies in the complex nature of environmental systems, while part lies in the social structures of knowledge building and information dissemination. These constraints "bound" decision processes (Simon, 1976).

Decision making involves the integration of three sets of primary components: available resources, information, and values. Many transactional decisions, such as the issuance of the septic system permits, have been bureaucratically structured to be almost entirely information determined. Standardized descriptions of the site characteristics dictate feasibility and design. Other decisions, such as the adoption of a zoning ordinance or the authorization of a capital project, may be primarily influenced by resource availability or values. These decision components are dynamic over time. The cumulative information available about Greenetowne's environment contributed to shifts in community values which in turn led to periodic adjustments in resources allocated to the environment. These extended processes are central to social learning.

For each decision, the availability and substance of relevant environmental information is highly variable. We can ask: what information is utilized by a prospective homeowner?; a firm selecting a new plant site?; a bank financing a manufacturer?; an insurer issuing a health policy?; an agency staff member reviewing a permit application?; a legislature adopting a new regulation?; a court reviewing a challenge to an impact statement?

It is useful to identify the various states of being informed. If being "informed" connotes the goal of having adequate environmental information for a rational decision, then "uninformed" represents the state in which the information exists but is not available. "Misinformed" means that incorrect information is used, while "disinformed" implies that the misinformation was intentionally provided by a second party. Finally, "ignorance" represents the condition in which inadequate information exists.

Our rational decision ideal, as embodied in NEPA, involves two distinct sets of information, descriptions of current environmental conditions, and forecasts or predictions of future conditions which would result from a given decision scenario. The development of current descriptions primarily involves data collection, classification, and mapping. The generation of forecasts and predictions is a branch of modelling, and utilizes methods ranging from statistical trend projections to mathematical simulation of transport paths.

## LIMITS TO SYSTEMS UNDERSTANDING

In Chapter 1, Karplus' spectrum of predictive knowledge was briefly introduced. The rationale for this continuum is based in systems theory. Modern science views the environment as a complex set of connected processes which are dynamically linked through the exchange of matter and energy. For example, Greenetowne's drinking water may be viewed from the perspective of the hydrologic cycle of Figure 2.1. Although widely embraced both academically and professionally for over a generation, this systems approach is only slowly becoming operational in decision making. One primary barrier is institutional: historically-established decision jurisdictions rarely coincide with natural systems' bounds. Greenetowne's Sole Source Aquifer underlies a crazy quilt of private, local, state, and Federal decision arenas. An equally challenging barrier is our endemic state of partial ignorance. For most environmental decisions, the development of a robust systems understanding inherently contains a number of important analytical challenges, including openness, reductionism, change, scale, and genius loci.

If we consider solar energy and its effects on climate as an example, it is easy to see that the entire environment is ultimately a single system. This universal truism has historically not been very productive because of the overwhelming complexity of the analytical challenge. The more typical managerial approach has been reductionist, identifying a specific geographic subsystem such as a watershed, or an important process such as a nutrient cycle. By artificially imposing subsystem boundaries, flows can be differentiated between "open" which cross the boundary, and "closed" which are internal. In reality; all environmental systems are to some degree open; this is a basis for deep ecology. Pragmatically, most analyses typically focus on the internal, closed flows, essentially assuming an isolated subsystem.

Since systems are dynamic, it is critical that a systems approach focus on change. Viable subsystems are presumed to have internal mechanisms that can maintain equilibrium over a range of external influences. A biological example is the cyclical shifting of a species' population corresponding to food availability, while an abiotic example is the change in a stream's erosion and sediment deposition patterns with fluctuations in energy related to water velocity. The systems approach includes two basic classes of change mechanisms: negative feedback, and positive feedback. In a physics example, pushing a child on a swing at the low point of the backward arc opposes the flow and slows down the ride; this is negative feedback. Pushing at the top of the arc accentuates the flow and increases the movement. This is positive feedback.

Many environmental systems maintain a rather steady state equilibrium over extended periods, as shown in Figure 4.1a. These self-regulatory mechanisms incorporate negative feedbacks which dampen or reverse change. A biological example would be the predator-prey population cycles of rabbits and foxes. At other times, in response to either internal or external processes, systems change states. This is depicted in Fig. 4.1b. This is often the result of positive feedback, in which the response mechanisms reinforce the direction of change. Examples include the eutrophication of a lake and a landslide induced by heavy rains (White, Mottershead et al., 1992).

Most current environmental regulations presume that the receiving system has an inherent carrying capacity. Beneath this limit the system can absorb external loading, such as pollution discharges, while maintaining equilibrium. If loading exceeds capacity, the system becomes unstable and may shift states. It is therefore of critical importance that decisions regarding environmental processes be based on longitudinal data to support our ability to understand change over time. This continuous learning is the key to knowing whether a system is functioning within an expected equilibrium range, or whether it is at the cusp of a major state change.

A third factor in developing a systems understanding is scale. It is generally acknowledged that environmental phenomena manifest different characteristics at different scales. In physics, we know that fundamentally different forces dominate the

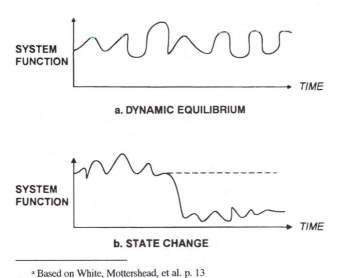

a. DYNAMIC EQUILIBRIUM

b. STATE CHANGE

[a] Based on White, Mottershead, et al. p. 13

**Figure 4.1. System States**[a]

subatomic, molecular, and gravitational scales. For the climate example, global warming influences but does not determine regional weather, which in turn is important to but quite different from the factors determining the microclimate on a forest floor that controls seed germination. Greenetowne's wetlands are composed internally of interacting organisms, water, nutrients, and energy processes. They also function as components of their local watersheds. In addition, they comprise an element of a continental flyway for migratory waterfowl.

Spatial scales are usually understood to be hierarchically linked. A comprehensive strategy for studying a system is therefore to identify a primary process and scale of analysis, and to study the relationships between it and the nesting scales above and below simultaneously (Allen & Starr, 1982).

A final challenge for the creation of systems-based knowledge is the uniqueness of place, the genius loci. It is essential for humans to impose simplified descriptive schema on our environment in order to reduce the infinite variety of data to a small understandable set of information. The most fundamental approach is the natural development of naming objects and phenomena which is universal in all languages. From its origins in the Enlightenment, modernism embarked on broad encyclopedic programs of identification and taxonomic classification. Such ordering has proven quite productive in that many environmental characteristics are generalizable. For example, a predominantly sandy soil based on grain size measurements will tend to be well drained, thus permitting rapid percolation of septic tank effluent.

Virtually all environmental analyses utilize multiple classification and measurement systems. The majority of these are reductionist, isolating one parameter at a time for analysis. Unlike manufactured materials in consumer products which undergo strict quality controls, however, in nature each specific site represents a somewhat unique synthesis of the history of its geologic, climatic, biotic, and human influences. This results in the genius loci, the individuality of place. The whole is much more than the total of its parts.

An outcome of this reality is that initial decisions based on environmental analyses are typically data- and knowledge-poor. The use of standardized classifications often misses local nuances. Budgets invariably limit detailed data collection, even on multimillion dollar Superfund cleanups. Since classifications such as climate, water, soil, and vegetation, are themselves reductionist, they rarely address the complex interactions between phenomena which have occurred over time in creating a site. The information challenges are both to improve the initial data, and more importantly, to generate continuing information so that subsequent decisions can provide the necessary midcourse positive or negative feedbacks necessary for our learning to nurture systems sustainability.

## THE KNOWLEDGE CONTINUUM

A systems view of the environment, with its associated analytical constraints, directly raises the question of how society learns about important phenomena such as

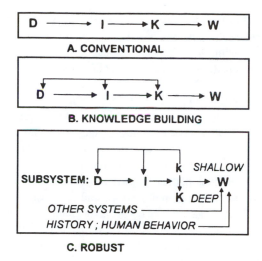

**Figure 4.2.  Knowledge Continuum**

those affecting Greenetowne. A popular simplified view of the learning process is depicted in Figure 4.2a, which linearly proceeds from data, to information, to knowledge, to wisdom. Briefly, data are represented by independent observations, such as traffic counts or water quality measurements. Data description typically utilizes one or more of four major measurement scales: nominal, ordinal, interval, or ratio. The naming of Greenetowne after Abigail Greene is a nominal assignment. The visual matching of a soil's color to "reddish brown" on a color spectrum is ordinal. The measurement that the river has dropped two feet from peak flood stage is interval data, while the counting of aircraft landings is ratio (note: both ratio and interval are numerical, but the former's having a known zero point allows a full set of numerical operations, including division).

Information is patterned data, such as the statistical mean temperature for a growing season, a map of critical habitats, or an annual graph of daily air pollution emissions for a factory. Processing often involves multiple stages of analysis. In contrast to just describing past or current environmental states, knowledge represents an understanding of a process or phenomena by which we can make forecasts or predictions regarding the future. The expected stream temperature next month is "x"; the air quality adjacent to the new road will probably be "y." Predictive capability is the central presumption of our environmental policies, from the broad sweep of NEPA to most of our media-specific emission regulations.

Clearly, the development of predictive knowledge does not proceed from the simple massing and analysis of randomly generated ad hoc data. Both science and technology utilize proactive cycles with an interchange between the development of knowledge about how a system works and the collection of data to confirm or refute the schema. The initial model or hypothesis may have come from primitive data such as observed seasonal cycles in water flow, or it may have come from a completely external concept, such as the analogy of gravitational attraction in traffic simulation of shopping trip destinations. In either case, the data collection and subsequent information processing is directly influenced by the current knowledge state as shown in Figure 4.2b.

As Karplus described, knowledge of an environmental process exists on a continuum from shallow to deep. In deep (white box) knowledge we have detailed understanding of the underlying causal phenomena. In an analysis of internal combustion and its associated emissions, science provides a fundamental understanding of the chemistry and physics involved. In contrast, a region which appears to have a higher than normal incidence of breast cancer typically represents a state of shallow knowledge. Statisticians may find a number of variables which exhibit positive correlations, but in general we do not understand either the multiple environmental sources, pathways, exposures, or the molecular, cell, and organ processes of disease generation. Figure 4.2c reflects this continuum. Shallow (black box) knowledge, as reflected in economic forecasts is primarily, statistical. Our understanding of most environmental processes is a mixture of shallow and deep, part causal and part correlative.

Wisdom involves the contextualizing of knowledge in making decisions. Again, the popular linear model of Figure 4.2a is naive in assuming that the sole basis of wisdom is more knowledge about the particular process in question. "We need more information/science before we can act" is a frequently heard conclusion to complex environmental decisions. In reality, wisdom transcends the narrow perspective of thematic predictive knowledge. Meta-knowledge is knowledge about knowledge, in particular its depth, quality, and scope. For the stream forecast and traffic prediction, do we really understand underlying causality or are we primarily analyzing correlations? What is the effect of data quality on error propagation? What is the cost-effective value of better information?

Meta-knowledge expands beyond the scope of the subsystem being analyzed to embrace both physical processes and social processes. Robust analysis considers the effects of other related systems. A wise decision incorporates systems complexity and lessons from history. It explicitly recognizes blunders, the potential for amorality, and shifts in values over time. In hindsight, it was unwise for Greenetowne to base its septic permit system on soil percolation, without considering the effect of waterborne substances on groundwater. It is unwise to set the allowable level of pollution emissions equal to the predicted "carrying capacity" of the receiving waters without a significant "factor of safety" to accommodate the fuzziness of both the performance prediction of the operation, and the complexity of the receiving system.

Similarly, it is unwise to base environmental quality judgements on industry self-reporting without considering both mis- and disinformation. Wise decisions necessarily consider the long run, and err on the side of flexibility and reversibility. This is the premise of Lee's social learning philosophy introduced in Chapter 1.

## MODELLING

In traditional societies, knowledge was commonly developed by long-term trial and error, with long periods of constancy punctuated by rapid breakthroughs. Egyptian irrigation, Mayan horticulture, Chinese fortifications, and Roman bridges all attest to the power of this vernacular approach to learning. (Note: there is little record of the efficiency of this approach since the vast number of failures are not recorded.) In modern science-based society, the generation of environmental information and knowledge has two facets, "real" and "symbolic". The ad hoc real world experience of the vernacular has been adapted to become either statistically-based data collection surveys, or controlled in situ or laboratory experiments. While this real grounding is essential to predictive accuracy, it serves as a complement to the dominant method of environmental analysis, symbolic modelling. Virtually all environmental management decisions incorporate one or more models, where a model is:

> A simplified representation of a system (or process or theory) intended to enhance our ability to understand, predict and possibly control the behavior of the system. (Neelamkavil, 1992, p. 30)

The classification of the Blue River's water quality, Greenetowne's zoning map, and the state's computer simulation of the aquifer pollution plume's path are all examples of models.

The field of modelling has spawned many taxonomic schemes intended to describe the variety of approaches that are available. Figure 4.3 depicts a simplified version for environmental models (White, Mottershead et al., 1992). The two primary

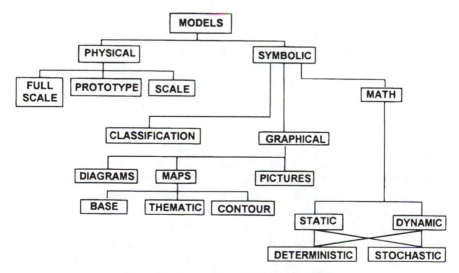

**Figure 4.3. Environmental Modelling Taxonomy**

groupings are Physical models, which involve the actual testing of physical elements, and Symbolic models, which entail the manipulation of representations of the environment. In practice, it is common to combine methods.

## PHYSICAL MODELS

Real world experimentation and testing of full-scale operations is a major component of physical modelling. For example, an incinerator may be given preliminary construction approval based on a computer simulation of expected air discharges. Final operating permission is dependent on successful test burns that are monitored at the stack and down wind. The test burn is a real world experiment. The meteorology and operating parameters during the brief test are a simplification of the range of conditions under which the facility will operate. This two step modelling-testing decision approach is an efficient learning process. It combines the flexibility and cost effectiveness of symbolic modelling with an explicit understanding of the limitations both in data and equations for waste generation, complex equipment, and meteorology, as well as the potential for human error (blunders and intentional) in manufacturing, construction, and operation.

Physical models also include prototypes and scale models. Prototypes are actual physical operating systems that are frequently of a smaller size than the projected project. An example of a prototype would be a new process to treat sewage. Development would typically proceed from basic science, the physics, chemistry, and biology of the underlying processes, to mathematical models of quantitative exchanges, to preliminary engineering of the system components. At this stage in design it is necessary to test the components of the system working together in a "realistic" context. A small prototype would be assembled at an existing treatment plant, and operated in parallel to the current process with a diverted fraction of the input flows. Evaluation would include both the performance of the system (often leading to cycles of redesign and retesting) and feasibility, the cost-effectiveness of the new option. Prototypes play a critical role in the continuing regulatory debate regarding "best available technology." When is a process proven and ready for certification, appropriate for recommendation, or inclusion in a regulation?

Scale models are small physical replicas that are typically developed to analyze selected subsets of environmental functionality. In scale modelling a replica is tested in a controlled context. These models have been used for thousands of years for visualization. Humans have an innate ability to perceive "bird's eye" views of three dimensional scenes. Such models are still popular today for communication of proposed land development projects to publics, elected officials, and potential investors.

Physical models are also used in making predictions. A dimensionally corrected model can be subjected to controlled conditions and the resulting effects measured. The development of a new high-rise building raises serious questions regarding wind forces in the structural design of the new facility, changes in stresses on existing buildings, and effects on pedestrian safety and amenity. The problem of analyzing urban wind is often too complex to be predicted mathematically, but can successfully be modelled in specially designed environmental wind tunnels.

Solar energy represents another field in which predictions have long played a central role in decision making. The regular patterns of solar and celestial positions have played a central role in both the cosmology and management of many historical societies. The Romans, as they simultaneously moved northward into Europe and depleted available firewood supplies, became acutely concerned with energy management. Through the regulation of city layout, building design, and innovative glazing they attempted to optimize the utilization of solar energy (Butti & Perlin, 1980). The "energy crises" of the 1970s in the U.S. led some states, such as California, to develop policies and regulations regarding both site solar access to promote use of solar collectors, and architectural or vegetative shading of windows to reduce air conditioning loads.

The location of shadow patterns at a particular time, date, and location (assuming no clouds) is a geometry problem and can be solved directly by math. Shadows may be also predicted using a scale model of the facilities and location by recreating the correct orientation of the model to the designated solar position. This may be done outdoors using a calibrated "gnomon," or indoors using an artificial light source contained in a "heliodon" apparatus (Felleman, 1990a).

## SYMBOLIC MODELS

Symbolic models encompass three major overlapping subgroups, Semantic, Graphic, and Mathematic. Semantics, the study of how meaning is conveyed through language, broadly ranges from vocabulary to syntax to rhetoric and the underlying logical constructions of contextual communications. The evolution of "swamp", a negative term connoting darkness, disease, and unproductivity, to "wetland", a environmentally positive (for some) value set, over the last 50 years reflects the vast importance of language in framing actors' decisions. "Ecology" has moved from an obscure sub-branch of biological science to a fuzzy but powerful set of concepts active in science, politics, and commerce.

For environmental information policy we are selectively interested in a few semantic-based concepts related to models, particularly classification. One of the major ongoing "projects" of modern science and environmental analysis is to describe and organize the environment and its phenomena in a manner that is meaningful and productive to various decision actors. From the earliest explorers through the proposed next generation of earth satellites, groups have been systematically labeling and mapping the environment.

A dominant traditional classification model has been hierarchial. The primary example is the assignment of all living organisms to a branch or twig of a taxonomic tree. This schema includes both a sequence of general to specific and an implied temporal continuum of evolutionary development.

Environmental hierarchies are not limited to biology. Basic environmental texts include world maps of climate and soils. These systems first combine a primary set of characteristics to categorize major groups. For example, temperature and precipitation yield "subtropical" climate. These groups can then be further divided into types such as "subtropical–dry summer." Environmental classifications are often hybrids. Because soils are the product of parent geology, and climatic and biologic processes, the classifications reflect these influences. For example, Browne County is found within the major group "middle latitude podzol" resulting from humid climates and mixed forest cover. Further subdivisions reflect finer and finer localizations. The U.S. Department of Agriculture's soil survey of Browne County had a two-level spatial hierarchy. General regional "associations" were delineated at a scale of 1:62,500, and local "series" delineated at 1:20,000.

Classification systems also play a major role in decisions related to socio-economic phenomena. The classification of land use and land cover is an input to most environmental decisions. However, numerous classification approaches for a phenomenon often exist simultaneously. In the Greenetowne scenario, each town has a different set of land use definitions in its zoning ordinance. The state transportation study uses a customized system focused on trip generation, the state tax system classifies on the basis of market value, while the USGS has promulgated a national approach which it has asked (unsuccessfully) states and locals to adopt and extend to larger scales (Thompson, 1988). A key information policy issue is the role of standardization in data generation.

While data standardization may be critical for data aggregation and the development of statistical information, overly simplistic classifications result in misinformation. Clearly, for a particular location, many descriptors may be needed to gain a comprehensive understanding of the land. These are discussed in Chapter 6. It is also important to note the inherent tendency in traditional taxonomic approaches to ignore systems hierarchy and genius loci. For example, in the USGS system, land is classified in a two-level hierarchy. Urban is major class (1) while residential is urban subclass (11). In this dendritic structure, it is not possible to distinguish the important local enclaves "forested residential" or "agricultural residential" because forests and agriculture are different major classes each having unique subclasses. A more robust classification approach, the design of the National Biological Survey, is discussed in Chapter 9.

All of the above classifications are the products of formal information policies. Once institutionalized, classification systems become primary resources both for their originally intended functions and for a host of downstream end users. The Standard Industrial Classification (SIC) is a key component of our national econometric analyses which study the interrelationships of various sectors of the economy. The SIC hierarchy includes: "Divisions," "Major Groups," "Groups," and "Industry." For example, Division D is Manufacturing, Major Group 22 is Paper and Allied Products, Industry Group 265 is Paperboard Containers and Boxes, and Industry 2656 is Sanitary Food Containers, Except Folding (OMB, 1987). When pollution regulations began to emerge it was convenient to utilize the SIC to help identify which enterprises were to be controlled. The Toxic Chemical Release Inventory (TRI) reporting requirements, discussed in Chapters 1 and 8, are targeted to SIC Codes 20 through 39. These codes constitute all of Division D.

Classifications that frame important environmental decisions are frequently controversial. Is the red wolf a "species" and thus covered under the Endangered Species Act? Is copper an "extremely hazardous substance?" Can (and should) the Census of Population and Housing ask people to classify themselves according to a short list of "races?" What combination of soil, water, and plants constitutes a "wetland?" In these examples the categories transcend purely descriptive functions and take on a significant normative role.

Many of these classification controversies, such as soils, land use, or wetlands, are complicated artifacts of the severe precomputer limitations on data analysis which went into the generation of a thematic map. A soils, wetland, or land use map represents a highly generalized singular synthesis of multiple data factors. In the manual production of maps the underlying data is lost to the end user, and often to the producer as well. This dead-ending of the information cycle severely limits the ability to learn. Geographic Information Systems (GIS), discussed below, can both retain data and provide multiple synthesis alternatives.

Most classification processing of environmental data to information involves a number of value judgements that have been standardized to achieve uniformity, efficiency, and communication effectiveness. Semantics are often combined with quantitative measures to derive the final classification. The latter is particularly significant when information users range from the highly technical to local officials to the citizenry at large.

## GRAPHIC MODELS

Since our perceptual systems are primarily visual, it is easy to understand the major role that visualizations play in environmental information. Proceeding from the realistic to the abstract, a continuum of graphic models includes: pictures and drawings, maps, and diagrams.

Sketches, photographs, video, and now computer visual simulations constitute a significant and growing fraction of our documentation and understanding of the environment. From field botanists using illustrated classification keys to identify plants, to lab technicians counting organisms under a microscope, to disaster relief workers videoing flood damage, to engineers developing designs for restoring a toxic waste site, pictures can uniquely capture and communicate superficial and internal environmental attributes and patterns. The explosion of digital technology, including satellites, scanners, digital cameras, and personal data assistants, has opened vast new uses for pictures and drawings to become a growing component of environmental information exchange.

Diagrams are typically utilized to visualize processes and relationships. Graphs of pollution emissions over time, and a chart of the stages for reviewing an environmental impact statement, are examples. Diagrams play a critical role in the developmental stages of both data flow analysis and mathematical environmental models. As discussed in Chapters 7 and 8, mass flow diagrams are becoming a key element of required pollution control reporting.

## Maps

The primary graphical model used in environmental information is the map. Since the beginning of recorded history maps have played an important role in describing the physical and social environments. The former includes the classification and location of selected characteristics such as soils and highways, while the latter addresses constructs such as political boundaries and property ownership. In Greenetowne, we saw a rich set of policy-related mapping products which resulted from science and regulatory and provision programs. These include: property surveys, zoning, soil survey, geologic survey, transportation design, site and building construction, and sole source aquifer delineation.

As models, maps necessarily incorporate a number of simplifications, many of which are a direct result of the historical format of the map as a specific physical product. These simplifications both facilitate information generation and communication, and limit knowledge building. From a policy perspective, the primary simplifications involve: projection, scale, map type, format/topology, and quality.

Each of the Greenetowne examples are "planimetric" maps, where "plan" represents the projection of features onto an idealized horizontal plane, and "metric" means that a uniform scale of measure is used in this plane. Thus, the distance measured on a planimetric map of a highway will be shorter than the odometer distance travelled because the map plane doesn't record the surface of the hills and valleys of the road profile. This has presented a major challenge to state and local transportation departments, whose legacy design and management information systems were based on actual odometer measures and are now converting to planimetric GIS.

Greenetowne's modern soil survey is published as planimetric mapping of soil units on an uncorrected airphoto base. This base is very useful in visually identifying ground features such as vegetation, roads, and buildings. The soils map includes a single map scale, but this is only approximate because air photos are filled with minor distortions due to airplane orientation, terrain relief, earth curvature, and lens parallax. In contrast, newer technology orthophotos contain all the rich information of airphotos with the elimination of distortion through photogrammetric processing.

Earth curvature presents another level of complication. The earth is a distorted spheroid. Numerous map projection systems have been devised to represent this three-dimensional reality in a two-dimensional map model. Each projection system uniquely slices and flattens a portion of the spheroid to create a coordinate framework for locating points, lines, and areas. An examination of the borders of a standard USGS 7 1/2 minute topographic quadrangle ("quad") map typically reveals three such systems and their associated units: Latitude/Longitude (degrees), Universal Transverse Mercator (UTM, meters), State Plane Coordinate (feet). Although most major governmental mapping agencies, national, state, and local, have formally adopted a projection system there is no universal standard.

EPA permits require facility locations in latitude and longitude. New York's Department of Environmental Conservation requires facilities to have a point location in N.Y. Transverse Mercator, while the state's Division of Equalization and Assessment requires ownership parcels to have points in State Plane Coordinates. In Wisconsin there coexists latitude and longitude, the original Wisconsin Transverse Mercator, the new National 1983 Datum Wisconsin Mercator, the Wisconsin County Coordinate System, and several local coordinate systems (Wisconsin, 1995).

Fairly precise conversion algorithms are routinely available for digital maps, but the accurate integration of maps from multiple sources is a fundamental information coordination challenge. Many GIS end users do spatial analyses that combine multiple maps and are unaware of the error propagation due to different coordinate systems. The national Metadata Standards discussed in Chapter 5 are designed to manage this issue.

## Map Scale

Map scale controls the size of the physical map product and the resolution of its features. Prior to the development of computerized maps, this was the dominant determinant of coverage and content. Map size is important for physical storage, display, and reproduction. Most jurisdictions have policies which standardize on sheet size, such as 24" x 36", for hard copy parcel, infrastructure, and regulatory information. In the digital world "seamless" maps can be created. For example, in New York, the Department of Environmental Conservation has released a CD that contains a composite of all the USGS quadrangle sheets for the state.

Map scale defines the reduction (representative fraction) between a distance in the real world and its distance on the map. The USGS 7.5 minute quad has a scale of 1:24,000, or 1"=2,000'. For a physical map a resolution of approximately 1/20–1/40" is a legibility limit, and this often determines the relative precision of delineation. For the quad, which is produced at National Map Accuracy Standards, the horizontal accuracy is approximately +/-50' on the ground. Important features smaller than the resolution are shown by symbol. For digital maps such as the new orthophotos, resolution is a more meaningful term than scale, since the computer software readily allows "zooming."

Professional guidelines exist for the appropriate selection of hard-copy mapping scale based on the intended use. These are summarized in Figure 4.4 (American Society of Civil Engineers, 1983).

Unfortunately, for many environmental decisions the available mapped information is at an inappropriate map scale:

> Resource thematic data such as soils and floodplains boundaries, are normally compiled at scales between 1:10,000 and 1:100,000 ... which may create erroneous information relating to specific parcels of land. (Committee on Geodesy, 1983, p. 104)

| APPLICATION | 1" = FT. | RATIO |
|---|---|---|
| CRITICAL DESIGN | 2–50 | 1:25–1:60 |
| GENERAL DESIGN | 40–200 | 1:500–1:2,500 |
| MICRO PLANNING | 100–1,000 | 1:1,250–1:12,500 |
| LOCAL PLANNING | 400–2,000 | 1:5,000–1:25,000 |
| REGIONAL PLANNING | 1,000–10,000 | 1:12,500–1:125,000 |

[a] based on A.S.C.E., 1983

**Figure 4.4. Map Scales and Uses[a]**

Flood insurance and wetland permits, for example, require a determination for each structure, a "general design" scale, while the official Federal Emergency Management Agency (FEMA), and Fish and Wildlife Service maps are similar to the USGS quads. The quads are the dominant set of maps used for the environment, and at 1"=2000' quadrangles are at the upper limit of "local planning" and best suited for "regional planning." The flood and wetlands programs are discussed in more detail in Chapters 8 and 9 respectively. Policies requiring the use of these maps due to their universality and low cost are a primary source of environmental misinformation to the landowner, lenders, and insurers.

A national study of state agencies prioritized the need for very large scale maps ("general design"). The study concluded that the generation of these maps is a state and local responsibility. The role of the Federal government was seen as primarily to develop uniform mapping standards. (Mapping Sciences Committee, 1990) The current national cooperative orthophoto program is helping bridge these gaps. The photography is designed to serve multiple national and state interests. Many states are going to use the corrected photos to develop digital base maps. States have the option to provide the additional resources needed to produce larger scale "quarter quad" products with digital pixel resolutions of 1 meter, and feature resolutions of 4–9 square meters.

## Map Types

If a map contains too much information it becomes impossible to understand, so a fundamental set of simplification decisions are made for each map. In the traditional paper map these decisions were absolute. The cartographer or map-maker went through a difficult set of tradeoffs involving audience needs, legibility, and production costs such as colors. Detailed input data that were eliminated or generalized were lost. In the digital world this restriction no longer exists. Data files can retain inputs while layers of related map information coexist in a GIS, allowing the ready development of customized map products. End users can interact with the spatial data to customize a display that meets their particular needs.

In environmental matters, four map types predominate: base, thematic, contour, and facility design. Most environmental studies and programs incorporate base maps. These serve two purposes: contextual orientation, and locational topology for subsequent subject maps such as thematic and contour (see below). The context locates the project or study area in its larger functional setting. For example, a location map for a proposed landfill should show the applicable political jurisdictions, the service area including primary haul routes, and the site location. Large projects often develop custom location maps for impact statements and other project documents. In contrast, many state and Federal environmental agencies require a standard use of the USGS quadrangle series for location.

The hard copy cartographer's decisions regarding base map construction have been widely relaxed in digital mapping. With the ability to superimpose digital layers, end users now have the capability to experiment with alternative base maps. Three emerging digital alternatives are the USGS' Digital Line Graph (DLG) version of the USGS quadrangles (Guptill, 1990), the Census' TIGER maps discussed in Chapter 9, and digital orthophotos.

For environmental phenomena which have mandated property owner responsibilities, such as wetlands, floodplains, and hazardous waste, the only appropriate base map is the tax map, which shows ownership bounds and is linked to the ownership and tax description database through a unique parcel number. A digital tax map superimposed on a digital orthophoto base maximizes content, legibility, and analysis capability. This combination represents a "deep" base map.

## Thematic Maps

A frequent map use is to depict the spatial distribution of a particular topic. In the 19th Century, two dominant models of environmental spatial distributions emerged, the patchwork quilt and the continuous three-dimensional surface. If we could climb to a viewpoint overlooking modern Greenetowne we would simultaneously see both models. The agricultural, suburban, and urban land uses tend to cluster in neat discrete patches which when classified and delineated produce a thematic map as shown in Figure 4.5a. In contrast, the terrain (both upland and submerged) and its associated elevations rises and falls continuously, generating a surface from an infinite number of point values. An efficient graphical representation of this phenomenon is a contour map as shown in Figure 4.5b.

In Greenetowne, examples of descriptive thematic maps included the geologic survey, the soil survey, and the wetlands maps, while the zoning map represents a prescriptive thematic map. In addition, many thematic maps can be generated from Browne County's census data, such as elderly by town, or census tract. In the soils example, a classification system is established, field data are gathered, and the mapping represents the best approximation of class spatial bounds. In the census example, the boundaries are determined prior to data collection. The primary analysis challenge is the selection of classes to depict.

Because thematic mapping, like all modelling, involves simplifying assumptions, the information conveyed is inherently limited. Traditional thematic mapping is both reductionist and static. The classification system forces all areas into one or another standard category (frequently a "mixed" category is included to provide some flexibility). From a systems perspective, the thematic map model has three fundamental problems: implied internal homogeneity, linear bounds, and static bounds (Felleman, 1992)

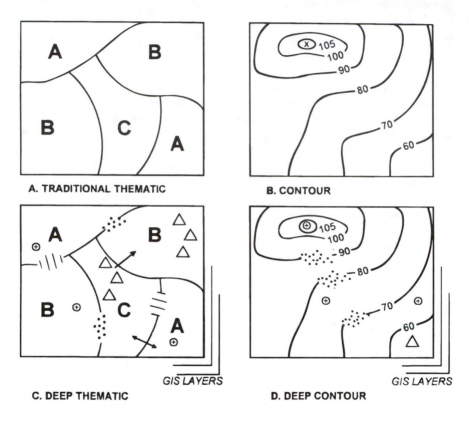

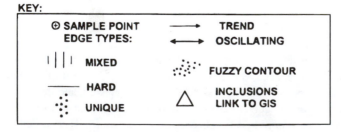

**Figure 4.5.  Map Types**

By convention, all of the area contained within a delineated boundary is assumed to be homogeneously composed of its labeled category. Thresholds are determined pragmatically by available resources. Typical county soil surveys have a size threshold. For example, they often do not map a unit smaller than three acres. Thus, homogeneous representations on the map may have inclusions of different soils. This is a prime reason why Greenetowne requires an on-site septic percolation test even though there exists a Federal soils map that specifically interprets "suitability for septic systems."

Frequently, as in the soil survey, the field staff and/or air photo interpreter actually can identify the inclusions but don't record them in the final mapping based on the thresholding rule. This rich genius loci data is systematically omitted. Where two or more categories coexist a decision rule, such as "select the category with largest surface area" is applied. In other situations the presence or absence of factors outside the theme being mapped may lead to the presumption that nonhomogeneity exists. In our soils example, elevation and slope may have played a significant geomorphic role in the development and transport of the soil and the development of inclusions. Although a GIS can connect two or more map themes to create a third, most standard environmental themes such as soils are not currently mapped in anticipation of subsequent GIS inclusion-probability modelling.

For over a century, interpreted boundary thematic maps have been universally depicted with single solid lines, as in Figure 4.6. Such an abrupt transition may occur, for example where a New England farm field bounded by a stone wall abuts a forest plot. Most edges are not so black and white. From a systems perspective, edges play a critical role in transport exchange processes. We can identify three basic type of edges. First is a hard edge, a clear, narrow line with abrupt differences on either side. This is the New England example and is the rule in traditional thematic maps. A different edge type is the transition zone. Here there is a region of blending between the adjacent cat-

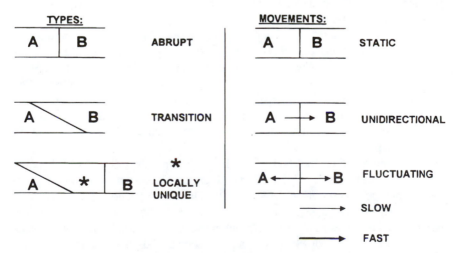

Figure 4.6. Spatial Edge Types and Movement

egories. Examples include the sediments at the edge of a stream, early vegetative succession along the edge of an abandoned farm field, and land uses at the periphery of a central business district. Finally, sometimes two adjacent process generate a unique microenvironment. These edge types are shown in Figure 4.6.

The third critique of the thematic map model is its static nature. In stark contrast to the rich fluxes of the actual environment, the boundaries of soils, land use, habitat, etc. maps are depicted as a frozen snapshot. Change, if depicted, is shown via a time series of static maps, each representing instantaneous states. Field personnel delineating thematic bounds or validating remote sensing interpretations can see various classes of dynamic edges. Some edges are essentially fixed for the period of the analysis. Others, such as erosional processes and some forms of suburbanization are essential,ly unidirectional. The edge moves in a discernible orientation over time. A third important type is the cyclical edge. For example, the boundaries of many wetlands move back and forth from year to year depending on the seasonal water level.

Figure 4.5c depicts a revision of the traditional conventions to include information quality and systems dynamics information. The result is a "deep" thematic map. Such a revised model is now technically feasible in an object-oriented GIS.

## Contour Maps

If the environmental phenomenon of interest is spatially pervasive and has a magnitude which changes in a generally continuous manner, then a contour map provides an effective first approximation model. In addition to depicting terrain, contour maps are the standard method of visualizing pollution distributions. A contour map is an information product interpolated from the underlying spatial data set. The quality and usefulness of the model is dependent on the spatial sampling which created the data set, the interpolation algorithm, and the fit of the contour model assumptions to the environmental attribute being studied.

Environmental sampling is both necessary and expensive. An air photo or satellite remote sensing scene represent continuous sampling. A variation is a regular systematic sample such as highway pavement conditions every 1/10 mile, or the USGS' Digital Elevation Models (DEM) which have a grid of values every 30 meters. Often the data values are not regularly spaced. Weather data are typically generated at airports; urban air quality is measured at major intersections; groundwater pollution data comes from monitoring wells that are located at expected critical positions in the plume's path.

Contouring, the interpolation of equal value isolines through the spatial framework of the data values, is a hybrid of art and science. Different assumptions in a contouring program will yield different results. The contour map typically assumes smooth continuous change. Many phenomena periodically exhibit dramatic abrupt changes. Temperatures near a weather front, population densities in areas with private and public lands, and bedrock geology may be poor candidates for standard contouring. Geologists, particularly in mineral exploration, have expanded the contouring programs to explicitly handle abrupt changes such as faults. Other disciplines have not yet followed this lead.

Direct contouring of raw environmental data may easily generate misinformation. Airport noise contours and groundwater pollution contours are two examples in which the only appropriate maps are the products of mathematical models, or professional judgement models, not the direct product of a contour interpolation algorithm. The transition between known data values is the resultant of physical properties and processes, not shallow statistical operations.

The basic contour map is a nineteenth-century naval invention. Comparing a topographic map with an evening news weather simulation shows the kind of evolution needed for environmental information. From a systems perspective, the desirable characteristics of a digitally-based contour map are depicted in Figure 4.5d, a "deep" contour map. Data values are posted. Smooth contours are interspersed with faults. Solid line contours become "fuzzy" where data/analysis quality is a probability function of data attributes in other GIS data layers. Expected directions of temporal change and magnitude are shown with a hierarchy of arrows.

## Facility Plans

The primary information format for the development and management of buildings, infrastructure, farms, and landscaping is facility site plans, which are done at a large design scale. These documents, which include drawings and maps plus their associated text contracts and specifications, are created for the parcel owners and managers. Facilities go through a common life cycle. Proposals are conceptual and schematic preliminary plans developed for obtaining financial backing. They are often the basis for environmental impact statements. Construction plans are the detailed documents used by contractors for building. They are also the basis for many specific governmental regulations, such as Greenetowne's building permits and Browne County's on-site septic permits.

Frequently, construction involves changes from the original design. Site conditions, such as bedrock, may differ from expected states, or owner financing or program needs may shift. "As-Built" drawings and maps record the actual construction product. Buildings in Greenetowne are inspected post-construction as a basis for a "Certificate Of Occupancy" as a check on the conformance of construction plans and the as-builts to the applicable codes.

Throughout the life of a site facility there are continuing cycles of maintenance, rehabilitation, and change. Farmers add drainage and irrigation systems; factories modify manufacturing processes to accommodate new products; homeowners build an addition. At times maintenance needs are neglected ("deferred") due to financial limitations and/or ignorance. At some point in time, an owner decision may be made to convert the site into a new land use. In extreme cases, such as bankruptcy, the owner abandons the facility. Site information generated during a facility's operation is highly diverse and primarily private. Knowledge building is complicated by sale. For exam-

ple, in New York State's approximately five million parcels, an average life of ownership is around seven years.

From a policy perspective, mapped facility plans are playing an increasingly
important role. Pollution cleanup, site audits, community notification, and pollution
prevention re-engineering are increasing the need for owners, both public and private,
to modernize and integrate their facility management information systems. Public
works, utilities, and government facilities such as military bases are aggressively converting paper maps to digital and integrating these into GIS and corporate database systems. A similar effort is under way in the manufacturing sector (Douglas, 1995).

## GEOGRAPHIC INFORMATION SYSTEMS

Until 20 years ago virtually all maps were "hard copy" physical products. The complexities of designing, printing, storing, updating, and reproducing physical maps
placed major constraints on their content, quality, and use. Digital mapping, which is
now seen as a component of comprehensive Computer Assisted Design, Geographic
and Land Information Systems (CAD/GIS/LIS) has revolutionized the field. Millions
of existing hard copy maps, such as utility locations and construction drawings, are
being converted to digital formats, while most new maps are being generated directly
as digital. Associated technologies such as satellite remote sensing and handheld
Global Positioning Systems (GPS) are now facilitating the direct digital capture of field
data and transfer to an automated mapping update system. Immediately following a
major flood, earthquake, or fire maps of the damaged area can be generated.

A crucial constituent of a digital map is its underlying topology. Two basic topologies, raster and vector, are the format for most mapping and GIS systems. A third,
object-based topology is currently emerging and is expected to dominate the next generation of systems.

The raster formatted map consists of a grid of data values. Each grid cell (picture
element, "pixel") represents a characteristic associated with the geographic space
located within the cell. This is shown in Figure 4.7(a). In essence a raster map is a form
of thematic map, in which an associated legend table connects the cell value with nominal or numerical data attributes. It is quite easy for a GIS to interact overlaying maps
to create new spatial models as depicted in Figure 4.7(b), because the location of cells
in each map dataset layer is identical.

In addition to map overlay, a raster GIS can readily handle a variety of spatial
modelling operations. These include: location of a data type; frequency of data types;
"buffering" (a distance search around a data type); cumulative data values along a path;
neighborhood analysis (such as slope); and reclassification. By combining these simple operations, highly complex models can be generated (Burrough, 1986).

Raster maps are directly compatible with raster data. Satellite remote sensing and
the USGS digital elevation models (DEMs) are two prime examples. Remote sensing
involves a plane- or satellite-based sensor which records passive (e.g. reflected light)

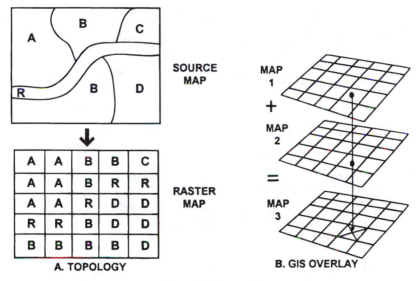

Figure 4.7. Raster GIS

or active (e.g. radar) energy intensity in a selected narrow band of the electromagnetic spectrum. Typically, more than one band is simultaneously sensed for each ground pixel. Subsequent classification of these data into a thematic map involves a combination of ground "truthing" with on-site data sampling, sophisticated statistical analyses, and simplification by smoothing and clustering adjacent pixels (Lillesand, 1994).

Square approximations of the environment are not suitable for many applications. Municipalities needed to analyze property boundaries and land use, watershed managers needed to overlay soil, slope, and vegetation maps, demographers wanted to associate population characteristics with census tracts and blockface street addresses. In order to address these needs, a more robust topology, the arc-node form of vector GIS, was developed. The Digital Line Graph (DLG) adopted by the USGS is a prime example. Its format and major associated data tables are shown in Figure 4.8. (Arnoff, 1989)

The spatial modelling power of the topology includes all of the basic capability described above for the raster model. In addition to maintaining the shape and precision of the input data, the arc-node approach has some unique capabilities, such as block face street address matching. Arc-node maps may be developed from hand-digitized physical base maps, by automated scanning and "vectorizing" of maps and other physical and digital images, and by direct digital input, such as a moving GPS. The potential for the visual scale of the arc-node output to misrepresent the precision of its inputs is much greater than for raster maps. In the latter, the texture of the pixels is at least a gross indicator of precision. In contrast, the arc-node systems have the ability to "clean up" images by smoothing lines, and snapping unmatched corners to create high-

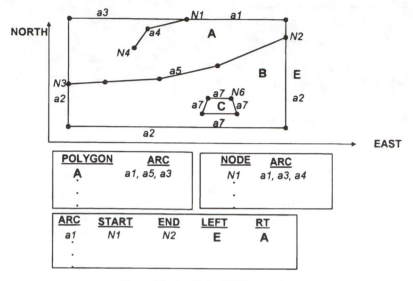

**Figure 4.8.  Arc-Node GIS Topography**

quality pictures regardless of the inputs. As discussed in the next chapter, the development of meta-data policy which includes map lineage and accuracy is critical to the proper use of this powerful technology.

Most major GIS software packages now include some mixture of raster and vector capabilities. Most also have proprietary algorithms for rapid, efficient implementation of these basic topologies. GIS information policy issues include transferability of digital maps between differing systems, and the reliability of GIS-generated spatial models. For example, in a comparative study of viewshed delineation (the mapping of terrain surfaces visible from a designated viewer position), each software system studied generated a significantly different viewshed map from the same DEM dataset (Felleman, 1990b). This was the result of the almost universal use of undocumented proprietary algorithms.

## MATHEMATICAL MODELS

The GIS systems discussed above are examples of hybrid graphic and math models. The disciplines of natural science and engineering, and more recently branches of economics and geography, are founded in large part on mathematical predictive modelling. A major and growing fraction of environmental analysis relates to mathematical model building and interpretation. From global warming to metropolitan air pollution, endangered species habitat needs, and superfund cleanup, mathematical computer models are the dominant predictive methodology. NEPA, in its logical positivist perspective, calls for predictions of impacts prior to decision making. In Greenetowne,

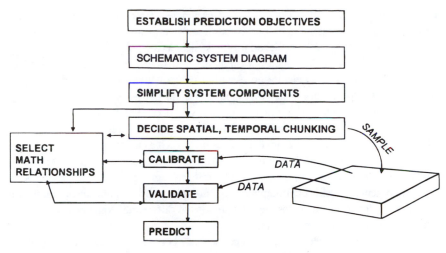

**Figure 4.9. Modelling Process**

regulatory processes require modelling for new air and water discharges, airport noise, traffic studies, and groundwater remediation. The arena of spatial modelling has been introduced above in the context of GIS. In this section the term "modelling" will be used to denote mathematical computerized predictive models.

A generic predictive modelling process is depicted in Figure 4.9. It is very information-intensive. The first step is to identify the objectives for the modelling, how the prediction will be used in decision making. Resource constraints, such as time and data availability, are established at the outset. The articulation of the target subsystem, its primary elements and relationships then leads to the formation of the schematic model. For example, if we are interested in predicting the peak runoff from a tributary of the Blue River, the subsystem is the terrain-determined watershed, and the schematic model is similar to the hydrologic cycle shown in Figure 2.1. Many simplifications are possible at this step. If the peak runoff is expected to follow relatively short, intense storms, the model may exclude evaporation, evapotranspiration, and groundwater outflow due to their small magnitude.

The development of the mathematical model involves a number of critical decisions. First, the modeler must "chunk" space and time. For the watershed example, the four-dimensional world is typically subdivided. In the simplest model, a single "lumped" parameter (P), such as mean slope, is directly estimated (high elevation - low elevation divided by length) for the entire watershed. A more sophisticated model uses distributed data to estimate a lumped parameter, such as average slope computed from local slopes. The distribution may be generated from a regular sample, such as a raster GIS, or from localized environmental subsystems, such as land-form sub-watersheds. By generating a lumped parameter, these models are necessarily limited to making predictions about the watershed as a whole.

The most analytically powerful approach to chunking models is to analyze processes at multiple scales. Here, data and parameters are established for each sub-unit. Predictions are generated for each subunit and energy and/or matter transfers and transforms are subsequently modelled for the system as a whole. The result is not only a prediction of flow at the outlet, but predictions of flows for all chunks modelled. Distributed parameter models are an essential component of ground water, surface water, and atmospheric analyses. For example, the current global climate models utilize atmospheric chunks 500 km on a side, 20 layers deep (Hileman, 1995).

The chunking of time in mathematical models falls into three general categories. The simplest is to predict for a single point in time, such as the peak flow. Such a simplification is a "static" model. Alternatively, "dynamic" models are designed to predict changes over time. Time may be broken into uniform segments such as hours, days, or population generations. Model outputs from this approach are like frames in a movie film. Finally, for some processes the mathematics of differential equations permits the analysis of time as a continuous function.

There is clearly a need for internal consistency between the space/time chunking and the selection of mathematical relationships. The model's equations are also a function of the state of our scientific understanding of the processes at work. As in Karplus' spectrum, some processes are well grounded in law-like behavior, while others can only be estimated empirically. In Browne County's ground water pollution model, equations describing the flow of polluted fluids through a uniform stratum of sand are mathematically determined from the basic physics of fluid dynamics and hydraulics. In contrast, for the County's traffic generation model, the commuter trips per zone equation involves a crude correlation of household demographics, income, and auto ownership.

There are numerous existing predictive models that are either generic, such as the Soil (now Natural Resource) Conservation Service's Urbanizing Small Watershed Runoff Model (McCuen, 1987), or have been developed for a specific location. In either case, it is desirable to adjust the model parameters to the specific location being analyzed. Normally this is a two-step process. First, using local data, the model's numerical factors are calibrated so that its predictions fit existing (or historical) observations. Next, using a second independent data set, the calibrated model is validated. If the validation results are not acceptable the model needs to be reconfigured. If results are validated, then a prediction is generated.

Our increased reliance on models has raised serious questions regarding their quality. Traditional models were "deterministic": the same data set always produced the same result. As discussed above, both the data and the equations may have quite a bit of internal "noise." Quality analysis methods have been developed to assess this problem. "Sensitivity analysis" is used to examine the influence of a single parameter on the results. The model is run repeatedly while the parameter in question is varied over its expected range. This is highly useful in making decisions regarding the cost-effectiveness of collecting additional data. In "stochastic" models all the major parameters are represented by distributions rather than constant values. Using Monte Carlo or other techniques, model predictions take the form of a distribution of expected values. EPA's recent Enhanced Stream Water Quality Model, QUAL2E, incorporates three types of uncertainty analyses: sensitivity analysis, first-order error analysis, and Monte Carlo simulation.

Since accuracy means closeness to a known standard, we can assess a model's accuracy regarding historical data at the stage of validation, but nothing meaningful can be said of the accuracy of a predicted future state at the time of the prediction since the future is not known. The quality analysis can assess the precision of the model. Ideally, the life of the model does not terminate at the time of the decision. Once chosen courses of action have been implemented, we can learn about the accuracy of the model by subsequent environmental monitoring. This feedback approach is central to knowledge building.

## SUMMARY

NEPA is a policy which attempts to protect and enhance the environment by mandating the incorporation of descriptive and predictive environmental information in Federal decision making. Subsequent legislation for air, water, hazardous waste, endangered species, and a broad environmental spectrum of issues shares a similar reliance on information. Recognizing the limits on existing information and models, NEPA and many of these acts also call for the development of new data and knowledge. As discussed above, our semantic, graphic, and mathematical models of the world are both powerful and limited. The positivism of NEPA has been tempered by a continuing series of "reality checks."

The revolutionary transformation from paper maps to GIS and related digital systems has vastly increased our ability to integrate and model spatial data. With this power has come the realization that the classification and mapping paradigms which have dominated the previous century are enfolded in existing data. Transforming this information to a digital format geometrically increases the possibilities for error propagation and the generation of misinformation. This is particularly problematic because most source maps do not contain adequate quality descriptions. GIS error propagation was a major research focus of the recent National Center for Geographic Information Analysis.

The usefulness and limits of information and predictive models can be seen in every arena. The cleanup of Superfund hazardous waste sites often involves the analysis of contamination plumes above and in the groundwater. Remediation efforts involve extended and expensive rounds of establishing liability and estimating expenses. Models are relied upon to assess various scenarios including the "null" (no action) case. The EPA and industry groups have wanted to standardize these models for the sake of efficiency and consistency. In an extensive review of this topic, the National Research Council concluded:

> Few flow and transport problems can be modelled with confidence... many of the models used in practice have not been validated... more uncertainty (comes) from the inability to precisely describe... parameters, the inherent randomness of geologic and hydrologic processes,... and measurement errors.... it would be unwise to rely solely on

any single source of information. Several agencies have guidelines that encourage the use of contaminant transport models. There are many different types of models, model applications, modelling objectives, and legal frameworks. Agencies cannot specify a list of government-approved models. (Water Science and Technology Board, 1990, pp. 5–8)

If this complexity exists in the "hard" physical sciences what about the use of models in biological or socio-economic arenas? Ecology was a specific focus in NEPA, and its integrative systems-process approach has been deeply enfolded in the modern environmental movement. This mixing of a scientific view and an ethical philosophy has frequently led to less than objective presentations of the strengths and limits of ecological predictive modelling. In an important review of this topic, Shrader-Frechette and McCoy find that "despite its heuristic power, general ecological theory has failed to provide a precise, predictive basis for sound environmental policy" (Shrader-Frechette & McCoy 1993, p. 3). They go on to identify a number of basic unresolved process problems such as community structure, carrying capacity, and biodiversity.

After a decade of millions of dollars of funding for large-scale, long term ecosystems studies, the International Biological Program (IBP) could provide no precise theories having predictive power. Hence the IBP provided little assistance to scientists who wished to use general ecological theories and their predictions to help justify specific environmental policies and actions. (p. 6)

The conceptual, theoretical, and evaluative problems associated with developing a precise, quantitative, and explanatory ecological science have suggested to some experts that ecology can play virtually no role in grounding environmental policy. (p. 8)

The authors disagree, somewhat. They are hopeful that there may be future progress in predictive modelling as scientific study continues. They propose that the appropriate current prescriptive use of ecosystem modelling is in developing practical advice in specific, locally data-rich, case studies. Even in these cases, which typically involve controversies of endangered species and critical habitats, the scientific information is necessarily enfolded in the ethical values of the publics involved.

Models that involve future human activities and natural processes play a significant policy role. Debate over the use and misuse of models in transportation decisions paralleled their emergence in the 1960s and has been an issue ever since (Lee, 1973). The Intermodal Surface Transportation Act and the Clean Air Act (1990) constitute a linked carrot and stick, project funding and regulatory control, approach to the improvement of metropolitan air quality. The EPA must certify that five-year and 20-year metropolitan Transportation Improvement Programs and Regional Transportation Plans, will meet mandated air quality standards. Projects and plans can include roadway construction, signalization and information improvements, transit, and management and regulatory programs such as emission controls, ride-share, zoning, and facility operations, including staggered work hours. The conformity decisions are based on predictions from linked demographic, land use, transportation, energy, and atmospheric pollution models.

The Denver region, with its rapid growth and non-attainment status, has long posed a problem with this modelling-decision process. The Denver Regional Council of Government has strongly criticized this process:

> The sheer complexity of the conformity requirements and the cumbersome and burdensome analytical processes do little to really improve air quality. The Denver region's nonattainment status requires 48 separate analytical tests- each of which must be passed to achieve conformity. The analytical requirements are extremely complex and rely exclusively on computer modelling to predict travel demand and resultant air emissions.

> Historically computer models used in transportation planning were designed to provide information about travel patterns and expected traffic conditions to assist policy makers in judgements about proposed facilities, information from such models typically has been weighed with numerous other factors as part of the decision process. Conformity, however, has forced a decision process that relies solely on numbers generated by transportation models not designed for this purpose; and air emission models that are little more than guesses about future air quality conditions. These models contain significant margins of error. Moreover, they are used to predict air quality conditions 10 and 20 years in the future... (this) results in multimillion dollar transportation project decisions hinging solely on analytical tools not intended for this purpose. (Denver, 1995)

From the standpoint of "deep information," modelling has a number of important implications. Properly done it constitutes a major tool for knowledge building, thus shifting our understanding from dark to light gray. This undertaking is data-intensive, with significant needs for shared and locally generated genius loci information. The models themselves are constructed with numerous simplifying value judgements. Explicit acknowledgement of these choices and pluralism in their selection can play a significant role in conflict resolution. Finally, as local governments, companies, and NGOs acquire increasing analytical capability, we can expect a rapid decentralization of model use. This will be fueled by the rapidly increasing availability of digital environmental data.

## REFERENCES

Allen, T., & T. Starr. *Hierarchy: Perspectives for Ecological Complexity*. Chicago: University of Chicago Press, 1982.

American Society of Civil Engineers. *Map Uses, Scales, and Accuracies for Engineering and Associated Purposes*. New York: The Society, 1983.

Arnoff, S. *Geographic Information Systems*. Ottawa: WDL Publications, 1989.

Burrough, P. A. *Principles of Geographical Information Systems for Land Resources Assessment*. Oxford: Clarendon, 1986.

Butti, K., & J. Perlin. *A Golden Thread: 2500 Years of Solar Architecture and Technology*. Palo Alto, CA: Cheshire Books/ Van Nostrand Reinhold, 1980.

Committee on Geodesy, National Research Council. *Procedures and Standards for a Multipurpose Cadastre*. Washington, D.C.: National Academy Press, 1983.

Denver Regional Council of Governments. *Transportation Conformity Issues*. Denver, CO: The Council, 1995.

Douglas, W. *Environmental GIS: Applications to Industrial Facilities*. Boca Raton, CA: Lewis Publishers, 1995.

Felleman, J. "Letters- Surfing or Drowning," *Geographical Information Systems* (Sept. 1992): 12,13.

Felleman, J. "Toward Smartsite." *Journal of Environmental Management*, 30 (1990a): 17–29.

Felleman, J. *Viewshed Determination: A Problem Analysis*. Syracuse, NY: S.U.N.Y.-E.S.F. Institute for Policy and Planning, 1990b.

Guptill, S. *An Enhanced Digital Line Graph Design*. USGS Circular 1048. Washington, D.C.: Government Printing Office, 1990.

Hileman, B. "Climate Observations Substantiate Global Warming Models." *Chemical Engineering News* (Nov. 27, 1995):/:8.

Lee, D. "Requiem for Large-Scale Models," *AIP Journal*(May 1993): 163–178.

Lillesand, T. *Remote Sensing*. New York, Wiley, 1994.

Mapping Sciences Committee, National Research Council. *Spatial Data Needs: The Future of the National Mapping Program*. Washington, D.C.: National Academy Press, 1990.

McCuen, R. *A Guide to Hydrologic Analysis Using SCS Methods*. Englewood Cliffs, NJ, Prentice Hall, 1987.

Neelamkavil, F. *Computer Simulation and Modelling*. New York: Wiley, 1992.

Office of Management and Budget. *Standard Industrial Classification Manual*. Washington, D.C.: NTIS, 1987.

Schrader-Frechette, K., & E. McCoy. *Method in Ecology: Strategies for Conservation*. New York: Cambridge University Press, 1993.

Simon, H. A. *Administrative Behavior*. 3rd ed. New York: The Free Press, 1976.

Thompson, M. *Maps for America: Cartographic Product of the U.S. Geological Survey*. Washington, D.C.: Government Printing Office.

Water Science and Technology Board, National Research Council. *Ground Water Models*. Washington, D.C.: National Academy Press, 1990.

Wisconsin Office of State Cartographer. "Too Many Coordinate Choices." *Wisconsin Mapping Bulletin,* **21** (2) (1995): 9.

White, I., D. Mottershead, et al. *Environmental Systems*. New York: Chapman Hall, 1992.

# 5

......................................................................................................................................................

# Information Policy

.........................

## INTRODUCTION

*I*n Section I, we introduced the emergence of our concern for sustainability from a historical perspective and, through the example of Greenetowne, saw the connections between information and decision making. Environmental information was first viewed at a macro scale as a data flow system involving environmental processes and a network of actors and stakeholders. Because sustainability is critically dependent upon social learning, we then engaged the more detailed perspective of the data-information-knowledge-wisdom continuum. Descriptive information of past and current conditions was differentiated from the use of models in creating predictive information for use in decisions.

For environmental information, the public sector has historically been the dominant arena. Because this sector primarily functions by means of policy creation and implementation, this first chapter of Section II will develop some basic frameworks for describing information policy.

In the Greenetowne scenario we saw a complex intertwining spiral of three threads: human spatial activities, a vital environmental system, and punctuated governmental interventions. The primary interfaces between these threads are physical and informational. The character and content of the exchanges across these interfaces is determined in large part by public information policy. Hernon has generalized:

> These policies have a profound impact on the political, economic, social and technological choices made by an individual, government or society. Policy issues may define individual freedoms in relationship to the good of society, the ability of government to carry out its functions, and the responsibilities and obligations of different stakeholders. (Hernon, 1989, p. 12)

An immediate challenge is that the topic is too broad to be understood as a whole. The creation and implementation of all public policy involves information generation and exchange. While some policy is aimed specifically at information processes per se, other policy deals with the vast spectrum of public concerns, including social and environmental issues.

There are certainly reasons for our difficulty in understanding information policies.... In spite of several attempts to define information as a resource... it remains an intangible enigma.... Political scientists themselves have a difficult time defining and understanding policy.... A third reason for information policy's elusiveness is the inescapable perception that it's involved with every choice we make. (Burger, 1993, p.5)

In this chapter we will introduce four constructs that will be useful in the discussion of environmental information policy: Public/Private; Formal/Informal; Target Arena; and Information Policy Lenses.

## PUBLIC AND PRIVATE POLICY

In our society, the fundamental unit of capital is land ownership. The land in Greenetowne is primarily privately owned and operated, with an intermixing of public facilities. Thousands of individuals and corporations and many independent governmental units continuously make decisions regarding their real property's life cycle. These decisions are the primary determinants of environmental quality and sustainability. Site location involves selection from alternative parcels and typically includes analysis of many variables, including first cost, access, improvements, services, amenities, liabilities, and taxes. Design and construction processes necessitate a balancing of programmatic functional needs with regulatory requirements and cost constraints. The daily operation of a home, farm, or factory involves numerous decisions regarding input, throughput, and outputs of materials and energy streams. Decisions regarding continuous, periodic, and/or delayed maintenance combine factors of performance, including regulatory infractions, and capital value decline reversibility with short- and long-term cash flows. The cycle reiterates as periodic decisions are made to rehabilitate, sell, or abandon.

At any point in time, most of the property in Greenetowne is not entirely controlled by the owner/operator of record. Rather, the vast majority has outstanding indebtedness consisting primarily of mortgages and liens. Many major commercial ventures are operated by managers and "owned" by shareholders. These investors increasingly have both financial and social interests in the operation of their facilities. In addition, virtually all of the businesses and most of the private structural improvements are insured for a variety of potential losses ranging from flood and fire to personnel health. Thus, the life-cycle decisions which affect the environment are often shared, multi-objective processes.

From the life cycle we can identify three primary sets of property actors: Governmental Regulators and Service Providers; Owner/Operators; and Lenders/Investors/Insurers. Each set uses internal policy mechanisms to structure its respective activities effectively. Governments create public policy, while individuals, organizations, and corporations have private policies.

In this book, the major emphasis is on public information policy. Adequate public environmental information is obviously necessary to inform the executive, legislative, and judicial branches effectively; it is a means to internal program management ends. Access to this internal public information is critically necessary for an informed citizenry to provide a check on the activities of government. Such internally oriented public information is, however, woefully insufficient for environmental sustainability. The land owners/operators and lenders/insurers are external end users of environmental information.

Over time, public and private policies are dynamically linked. For example, hazardous waste cleanup laws have dramatically changed the criteria banks use for making loans, while simultaneously stimulating the lending community to demand better public site-specific historical pollution information. This topic, discussed in Chapter 8, illustrates a publicly facilitated market-based stewardship approach. A classic criterion for a viable market is the existence of well-informed buyers and sellers. The traditional economic explanation for environmental degradation is that it is an "externality." The despoiler was not directly accountable, typically because of public or market ignorance. Elimination of the externality "market failure" problem constitutes a basic justification for government involvement.

Public information policy plays a significant role in the move toward environmental sustainability by meeting the information needs of the private marketplace actors. However, the needs of these external end users' transcend just having access to existing internal governmental information, a policy approach often called "throwing candy (or garbage) over the fence."

## FORMAL AND INFORMAL POLICY

There are as many definitions of "public policy" as there are texts on government and political science. This spectrum continues to grow as the complexity of our society increases, the diversity of governmental activities expands, and our understanding of social systems dynamics deepens. A simple rationalist approach to defining policy is reflected by Jenkins:" (policies are)… decisions… concerning the selection of goals and the means of achieving them"(Burger, 1993, p. 7). From this perspective, each of the four major branches has designated responsibility for specific types of policy promulgation: Legislature: law making; Chief Executive: executive orders, administrative standards; Agencies: program rules and regulations; and Judicial: decisions reviewing actions of the other branches in the context of the Constitution.

Such policies are "formal." Each formal decision is the product of a specified process, and is written in the public record. Much of the study of public policy focuses on the creation of formal policy, due both to its mandatory predominance and its relative ease of accessibility. The constitutional construct of separation of powers to provide a system of checks and balances has led to a government of multiple rationalities. The process of policy formation differs significantly between branches. Chief Executive and Legislative policy making is the study domain of political science policy research, which tends to emphasize the dynamic game-like sequences of trading and coalition building. A more rationalist approach to achieving efficiency and effectiveness is taken in the domain of public administration study of the agencies. Due to their small size and high visibility, policy making by the Chief Executive and high courts appear to share a high level of individual determinism by key actors.

In contrast to formal policy making, since Simon's pioneering work in analyzing organizations we have known that a significant portion of decisions made by agencies are informal (Simon, 1976). These decisions are only partly rational ("bounded rationalism"), may not follow a particular process of evaluating alternatives in light of specific goal achievement, and typically reflect the value sets of internal subcultures.

A definition of policy which encompasses this broader set of decisions can be found in Jones, quoting Eulau and Prewitt:

> Policy is defined as a "standing decision" characterized by behavioral consistency and repetitiveness on the part of both those who make it and those who abide by it. (Jones, 1984, p. 26)

From this perspective, government inaction may also constitute policy. For example, there are formal policies, laws and regulations, which cover the granting of permits for the construction and operation of hazardous waste facilities. In recent years, a growing set of stakeholders has coined the phrase "environmental racism" to focus public attention on the endemic pattern of environmental health problems being disproportionately associated with poor, minority communities (Foster, 1993).

Does the Federal government have an "environmentally racist policy?" The term "racism" is highly electrifying and will not be engaged here, but the underlying issue of "environmental justice" is certainly valid for policy discussion. As one might expect, none of the formal policies make an explicit negative reference to income or race. There is, however, also little argument that market forces will tend to locate facilities in the most cost-effective sites, and lower income districts tend to have lower land prices. In addition to initial site location, pollution enforcement is clearly a tactical decision involving scarce resources and political pressures. It's to be expected that regions with under-empowered publics may have an above-average concentration of regulatory violations.

The "de jure-de facto" dimension of this formal-informal policy issue is just the most recent example of a long series that includes Native American displacement, coal mine safety, strip mine acid runoff, National Forest over-harvesting, and barrio construction on the U.S. side of the Mexican border. Using Eulau and Prewitt's definition, a "de facto" argument can be made that a consistent pattern over time of regulator–regulatee interaction or agency program (in)activity constitutes public policy. From the standpoint of ecosystems and natural processes, it is the physical activity, not the social rationale, which matters. Any serious analysis of environmental policy must therefore transcend the formal statements of intent and include a comprehensive reality check in the physical world. Valid information plays a central role in the policy check and balance between formal and informal policies.

In 1990, the EPA administrator established its Environmental Equity Office. A primary objective of this office is to:

establish and maintain information which provides an objective basis for assessment of risks by income and race, beginning with a research and data collection plan. (Environmental Equity Office, 1990)

In 1993, a presidential Executive Order was issued on "Federal Actions to Address Environmental Justice in Minority Populations and Low Income Populations." Seventeen Federal agencies are involved in the directive, aimed primarily at the equal enforcement of regulations.

## POLICY PROGRAM TARGET ARENAS

In addition to the dichotomy of formal/informal, another useful schema in understanding public policy focuses on the program target arena. The major arenas are programs and administration. The basic function of the Executive Branch is to implement the law which has been created by the legislature and/or interpreted by the courts. In the Greenetowne example, the most visible set of governmental activities involve programs. Regarding the environment, two primary program subarenas are "control" and "provision." Examples of controls are regulations of the private sector such as land use zoning, and regulation of the private and public sectors such as surface water pollution discharges. Some controls, such as zoning, are implemented at the level of government legislating the control. In contrast, many pollution regulations originate at the Federal level, and are primarily implemented through intergovernmental transfers to the states. The information policy dimensions of regulatory programs are the focus of Chapter 7.

Provision program arenas include: direct government projects, such as parks and sewage treatment; intergovernmental transfers such as the Federal highway system, which is implemented by the states; resource transfers such as pollution control tax credits; and information/education such as the Soil Survey and agricultural erosion control planning. As Greenetowne clearly illustrated, over time the environment is cumulatively affected by numerous, often disjointed service programs. Chapter 6 includes an analysis of the central role of public land records, the cadastre, while Chapter 9 provides an critical review of several science information provision programs.

## ADMINISTRATIVE INFORMATION POLICY LENSES

While program policy deals with "what" the Executive Branch does, administrative policy deals with "how" the agencies operate. A major reference to this field is *United States Government Information Policies* (McClure, Hernon & Relyen, 1989).

Following a controversial half-century of rapid, uncoordinated Federal agency growth, the comprehensive Administrative Procedures Act was passed in 1946 (5 USC

551). In the next decade a model State Uniform Procedures Act was promulgated, and today most states have adopted a standardized approach to adjudication, rule making, and judicial review. Administrative policies encompass both the internal activities of an administrative agency and its interactions with client groups and the public at large. An example of the former are the Federal Information Policy Standards (FIPS), which control the format and processing of Federal records. An example of the latter is the Freedom of Information Act (FOIA, 5 USC 552) which regulates how citizens may access Federal records.

It is important to view the major environmental policy program arenas, control and provision, in the context of the administrative policies in which the agencies exist. Hernon has stated:

> Information policy is based on a set of interrelated laws and guidelines that govern the information transfer process (Hernon, 1989, p. 12)

What he is primarily referring to here is administrative information policy, which necessarily focusses on access as a fundamental component of our system of checks and balances. Henron subsequently identifies three related primary administrative information policy issues:

1.  Production and maintenance of information;
2.  Dissemination of information; and
3.  Information relationships between stakeholders.

For environmental information policy it is useful to expand these policy arenas to embrace both administrative and program targets. Therefore a rearticulation of Hernon's

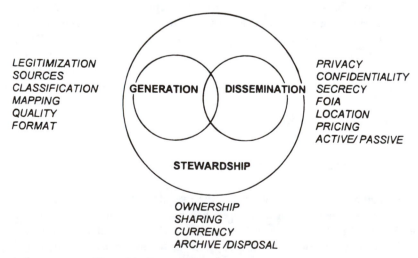

LEGITIMIZATION
SOURCES
CLASSIFICATION
MAPPING
QUALITY
FORMAT

GENERATION   DISSEMINATION

PRIVACY
CONFIDENTIALITY
SECRECY
FOIA
LOCATION
PRICING
ACTIVE/ PASSIVE

STEWARDSHIP

OWNERSHIP
SHARING
CURRENCY
ARCHIVE /DISPOSAL

**Figure 5.1.  Environmental Information Policy Lenses**

categories results in the policy frameworks depicted in Figure 5.1. These lenses include Generation, Dissemination, and Stewardship. Each has a number of facets.

## INFORMATION GENERATION POLICY

How do governments obtain data and transform them into environmental information? This simple question leads directly to the highly dynamic policy arena of information generation. Policy facets include: Legitimization, Sources, Classification, Mapping, Quality, and Format.

### Legitimization

Virtually all public environmental information efforts are initially legitimized by a legislative or agency program directive. The most striking example is the requirement in the U.S. Constitution for a decennial census of population discussed in Chapter 9. Information legitimization in legislation runs the gamut from a specific call for particular data, to more general guidelines for reporting. Examples of the former are the listing of chemicals in pollution laws, and the requirement for the creation of flood maps, while the latter includes NEPA's description of Impact Statements, the mandate for agencies to report to congress such as the Natural Resources Inventory, and the effectiveness of the Pollution Prevention Act reporting.

Typically the legislation is of medium focus, thus passing the responsibility to the bureaucracies to develop the specifics of information generation programs. Some information programs are the result of broad agency responsibilities. The Department of Agriculture, the U.S. Geological Survey, the Weather Bureau, and the National Biological Service were all created to advance our understanding of environmental processes. Agency heads often have wide latitude in interpreting and initiating information generation process. A recent example is the Interior Secretary's internal reorganization and start-up of the National Biological Survey.

For some information management efforts, like EPA's initiative to integrate its separate facility databases, information processing decisions are essentially internal to the agency. In contrast, regulatory information requirements involving the private sector are legitimized by the steps of the Administrative Procedures Act (5 *USC* 551). These include: public notice, comments and/or hearings, and final notice before implementation. In recent years the comment period has often been expanded into negotiated rulemaking with major stakeholders.

### Sources

Governments utilize a variety of sources to obtain environmental data. As depicted in Figure 5.2, these include: (a) Direct Governmental Data Collection, "first party"; (b) Mandatory Private Sector Reporting to Government, "second party"; (c) Mandatory

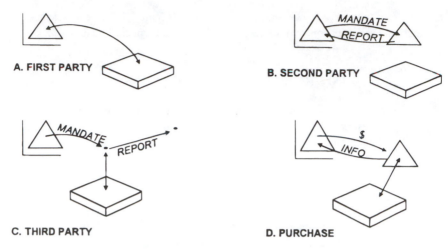

Figure 5.2.  Government Data Generation

Reporting to a "third party," and (d) purchase of private sector information. Each approach has unique strengths and weaknesses regarding operational efficiencies and information quality effectiveness.

First party generation entails direct government data acquisition and processing. From Lewis and Clark's expeditions to proposed earth monitoring satellites, most of our environmental science programs are in this category. These are discussed in Chapter 8. A fundamental first party filter issue involves the public acquisition of environmental data from private property. Three collection modes are remote sensing; in-situ outdoors; and in-situ indoors.

In general, there are no restrictions regarding remotely sensed data, such as photography and thermal or infrared detection. The old British common law principle of "open fields", wherein what could be seen on a farm or estate from a public road was allowable evidence, has been extended by U.S. courts to include satellite and airplane sensors. The actual use of remote acquisition varies, depending on community values. For example, many local governments routinely utilize low level oblique color photography to identify building violations such as unpermitted additions and swimming pools. In the Adirondack Park, the informal policy of the Park Agency Commissioners Board has been to ban to use air photos to detect violations because of the outrage toward "big brother" spying voiced by numerous local landowners.

In contrast to remote sensing, the Constitution does explicitly protect private property, and the government needs specific regulatory or judicial authorization to enter onto a parcel or inside a building or mine. Again, de jure legitimization does not necessarily mean the data will be collected. Numerous legitimate environmental generation activities, ranging from local tax assessments to Federal Occupational Safety and Health Administration (OSHA) inspections, are curtailed due to local de facto resistance and the resulting political pressure on the agency to desist.

A common variation of first party generation is the federalism model, wherein a higher unit of government delegates or requires data generation to a lower unit. Data about the Federal highway system, and the collection of regional air pollution information, are examples.

A second variation of first party data generation entails the legitimizing of citizens to collect the on-site information for the government. Legitimizing includes a formal policy of delegation, training related to collection and reporting standards, and monitoring quality control. Typically the government legitimizes an NGO rather than dealing directly with individual citizens. While the initial motivation may be to save scarce public resources, the empowerment of locals to systematically observe, report, and affect public actions regarding their vested environments is a direct and powerful engagement of the activist modern community which Dewey had championed in the early part of the century. Dewey's general call for community learning was first focused directly on the environment in the nineteen thirties by L. Mumford for the pioneering New York Regional Plan.

Due to the filters discussed above, the use of citizen data collectors is usually limited to publicly accessible sites such as navigable waters, public land, and hiking trails. A primary example of legitimized NGO-public data generation is the monitoring done by hundreds of "lake associations" throughout glaciated North America. Primarily composed of cottage owners and fisherpeople, these groups provide an unmatchable source of spatial and temporal water sampling and pollution incident monitoring. These local groups are sanctioned by state environmental agencies. In 1986, New York's Department of Environmental Conservation started its Citizens' Statewide Lake Assessment Program. The data flow diagram for this public-private partnership is shown in Figure 5.3 (N.Y. Department of Environmental Conservation, 1992). On a much larger scale, the Lake Superior Watch program

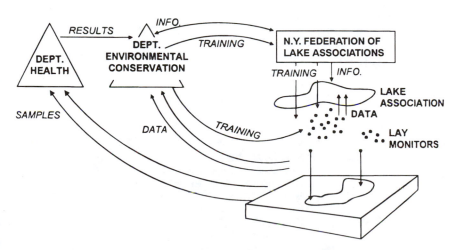

**Figure 5.3. NY Citizens' Statewide Lake Assessment Program**

includes citizen monitors from three states and the Province of Ontario.

The Texas Watch program is a state-wide coordinated effort that includes both surface waters and well head protection areas. The latter is a primary focus of local Retired Senior Volunteer Programs. The Texas Watch quality assurance program has been approved by the EPA, allowing its data to be included with governmental records in analysis, planning, and regulatory processes.

The EPA now actively solicits standardized group monitoring programs. This approach of citizen data generation transcends water quality issues. The North Carolina Environmental Defense Fund uses citizen monitors to survey ozone injury to plants in wilderness areas and national forests of the Appalachians. This bio-indicator data is intended to complement the government's ongoing but limited point sample instrument-based measures of ozone.

*Second Party.*   One of the main sources of public environmental data is mandated private sector self-reporting. The U.S. census of population, called for in the Constitution, is the largest and most thoroughly analyzed program in this category. Most second party data are generated by the regulatory programs. The private sector describes itself in permit applications, impact statements, and conformance monitoring. These programs are the subject of Chapter 7. Needless to say, issues of missing data and disinformation are endemic.

A major generation policy filter for second party data is the Paperwork Reduction Act (44 *USC* 35) and its amendments (P.L. 96-511). It establishes the Office of Information and Regulatory Affairs in OMB which must approve all agency proposals for the new collection of information from parties outside the government. This is shown in Figure 5.4. Included in OIRA's analysis is the performance based-need for the information, an "objectively supported estimate of burden" on the public.

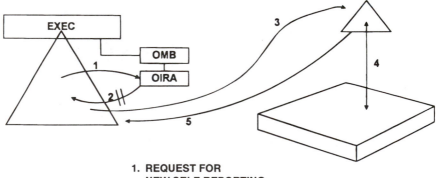

1. REQUEST FOR
   NEW SELF-REPORTING
2. BENEFIT/COST BASED APPROVAL/
   MODIFICATION
3. REPORTING MANDATE
4. DATA COLLECTION
5. SELF-REPORTING

**Figure 5.4.  Paperwork Reduction Act Second Party Filter**

A second thrust of this act is to reduce the burdens of redundant information. For example, it is not unusual for a major manufacturing facility to have 20 or more simultaneous environmental permits. A large fraction of the required data on each form which describes the facility is repeated. The objective of "clarification or simplification" relates both to within a single data request, and between multiple data requests.

*Third Party.*   Government in the United States has always been reluctant to get directly involved in private sector transactions due to the contract clause of the Constitution and to a general dislike of meddling by the citizenry at large. One area where this has historically been relaxed has been in the sale of real property. This is the topic of Chapter 6.

Information has always been both a complement and an alternative to regulation. In recent years, as our understanding of environmental complexity has grown, the potential for an information-based approach to stewardship has become more desirable. Major precedents have been established in the field of consumer protection, such as labelling and "lemon" laws. Third party generation involves the governmental mandating of owners to notify buyers of a potential environmental problem, as shown above in Figure 5.2(c). Two examples involve California earthquakes, and Florida radon.

The state of California has earthquake provisions in its building code, and it has a statewide earthquake mapping program at a scale of $1 = 2000'$ which depicts seismic zones of three hazard types. Earthquake safety involves the structure, the interior decoration, and the occupants' lifestyle. The California law states that at the time of sale the seller must notify the buyer of the parcels' location in a mapped seismic zone (Real & Yoha, 1991). Radon will be discussed in Chapter 8.

Third party information policies are increasingly popular ways for government to address spatially dispersed issues of environmental stewardship. As a policy option, it involves low governmental cost, and minimal intrusiveness. It remains to be seen whether these approaches are substantive or symbolic. They do not result in coordinated data generation, are not quality controlled, and are a questionable basis for social learning.

*Purchase.*   Many data sets used by the government are purchased from the private sector. Some of these, such as the business information generated by Dun and Bradstreet, are widely marketed commercial products. Other data, for example a special low-altitude remote sensing flight covering a fire in a national forest, are custom contract products.

The increasing pressure for privatization of public programs will continue the growth of this source. Policy issues regarding information privatization center on ownership, access, and pricing of the datasets for external endusers.

## DISSEMINATION

Two opposing principles have interacted to help shape the continuing evolution of public information dissemination. On the one hand is the position of many founding fathers, particularly Jefferson, that democracy requires a well–informed citizenry. This was originally interpreted as a component of accountability checks and balances. The government was "of the people." In order to vote effectively, the electorate needed to understand how government was being conducted. A second component of informed citizenry, including business, emerged after the Civil War in response to the accelerated growth of science and technology. Here the justification was economic. Government concluded that science, education, and technology transfer was a public good. The establishment of land-grant agricultural colleges and networks of extension agents is a prime example.

On the other hand is Weber's conclusion that a fundamental operational characteristic of bureaucracies is the desire for secrecy. This insulates the organization and provides a basis for relative power (Gerth & Mills, 1946). These opposing positions, informed publics and agency secrecy, generate a continuing policy dynamic which has become highly active as we continue to automate our information resources.

### Active/Passive

Information dissemination policy has a number of important facets, including: passive/active access, privacy, confidentiality, location, secrecy, and pricing. Relyea has developed a succinct overview of the benchmarks in the evolution of our Federal information policy regarding access and dissemination (Relyea, 1989). The growth of big government, starting in the depression, and continuing in World War II and the post-war economic boom has severely stressed Jefferson's goal of an informed citizenry.

Roosevelt's explosive, sprawling New Deal agency growth frequently left citizens and companies in the dark regarding both service and regulatory programs. This was particularly critical for the latter where the legislature frequently delegated a combination of legislative (rulemaking), judicial (adjudication), and administrative functions. Reform studies which were started in the Depression ultimately led to the passage of the Administrative Procedures Act in 1946. Much of this act focused on standardizing the processes of rule making. Regarding agency information, the APA:

> contained an important public information section which directed the agencies to publish in the *Federal Register* 'the established places at which, and methods whereby, the public may secure information or make submittals or requests'. Unfortunately, broad discretionary allowances also were made for protecting information. (p. 35.)

The result was that information was available on a "need to know" basis, with the onus on the requester to justify the need. There was no direct legal recourse if information was denied.

This approach was amplified by the Cold War and its pressures for secrecy. However, the continuing rapid growth of government as we entered the "Great Society" period led to a groundswell of pressure to reform accessibility to Federal information. The FOIA created a presumptive "right to know" with explicit public access to administrative records. It is now the government's burden to justify under one of FOIA's nine exemptions why access is denied. Many of these filters directly affect environmental information.

Under Exemption 1, documents may be "classified" in the interest of "national security." Most military bases and weapons development and storage areas fall in this category. These sites may represent the largest concentration of toxins and hazardous waste in the country. Exemption 4 excludes access to "confidential business information," including trade secrets and business plans. These will be discussed in Chapter 7 in relation to Community Right-to-Know, and the Pollution Prevention Act.

Exemption 5 covers "internal government communications" which precede a final action. This is necessary to facilitate the candid exchange of ideas regarding alternatives. The result is that, for most intra- and inter-agency responses in Environmental Impact Statements that the public sees, the original technical and scientific reviews have been edited by administrators to reflect executive positions. Exemption 6 is for "personal privacy." This has been reinforced by the Privacy Act of 1974. The ramifications of this filter are frequently a function of the details of database design. For example, an analyst developing a model of non-point pollution in a watershed may be denied copies of the Department of Agriculture's farm management records because the field data are in the same flat file as data about crop support finances.

Exemption 7 covers information that is actively involved in an enforcement proceedings. Thus, a community group often cannot access government data regarding a local polluter. Finally, Exemption 9 deals with geological information about oil and gas wells. The government needs information about reserves in order to do strategic planning, while the energy companies need confidentiality in a highly competitive market. Although this information may be important for the analysis of pollution and ground water aquifers, it is not available to the public (Committee on Government Operations, 1991).

FOIA is the last major macro information policy of the precomputer era. The records it covers and the procedures for public access are primarily based on a hard paper copy, early Xerox information technology model. As noted above, the Paperwork Reduction Act of 1980 centralized Federal information policy in OMB's OIRA and focused on the transformation to electronic technology. The Act calls for the establishment of the Federal Information Locator System to include data profiles of major information holdings. These profiles were to include directories of information holdings; data dictionaries; and information referral services.

As Federal agencies rapidly adopted new systems, it became readily apparent that FOIA needed a major rearticulation. In a landmark study, the Office of Technology Assessment reviewed the traditional dissemination channels, including the Government Printing Office, the National Technical Information Service, the Congressional Record, and Federal depository libraries as well as FOIA, and identified the need for sweeping policy and program transformations (Office of Technology Assessment, 1988).

FOIA was not legislatively revised at this time due in part to debates over privatization and downsizing of government service programs.

In 1993, OMB, under the authority of the Paperwork Reduction Act, promulgated a major revision to previous information policies. Circular No. A-130 requires Federal agencies to implement a comprehensive Information Resource Management plan which emphasizes electronic information. Agencies are called upon to cooperate in their management of information resources and to consult with state and local governments in the development of policies and programs.

> The original Circular No. A-130 distinguished between the terms 'access to information' and 'dissemination of information' in order to separate statutory requirements from policy considerations. The first term means giving members of the public, at their request, information to which they are entitled by a law such as the FOIA. The latter means actively distributing information to the public at the initiative of the agency... popular usage and evolving technology have blurred differences between the terms.... For example,... online 'access'... becomes a form of 'dissemination'. Thus the revision (to A-130) defines only the term 'dissemination'. (Office of Management and Budget, 1992)

Federal policy, both the revisions to the Paperwork Reduction Act (44 USC 35) and the President's National Performance Review ("reinventing government") program has embraced Government Information Locator Services (GILS) as the primary mode of information dissemination. In addition to a wide area network environment, the locators require standardized metadata both for the internal indexing and to assist the end user in deciding whether the available information is suitable. The leading environmental model is the National Spatial Data Infrastructure (NSDI) coordinated by the Federal Geographic Data Committee (FGDC).

The origins of this effort began in the 1980s when OMB reacted negatively to the multiple, independent requests by Federal agencies to develop new internal GISs including massive new data sets. The initial reaction to sharing was both an ignorance of existing efforts and a mistrust of data quality. After years of informational meetings and inventories, OMB issued Circular No. A-16 which mandates coordination in:

> surveying, mapping, and spatial data activities.... A major objective of this Circular is the eventual development of a national digital spatial information resource, with the involvement of Federal, State, and local governments, and the private sector. (Office of management and Budget, 1990)

The FGDC has made excellent progress toward this goal. A Content Standard for Geospatial Metadata was approved in 1994. Metadata includes standardized information about: identification, data quality, spatial organization, spatial reference, entity attribute, distribution, metadata reference, citation, time period, and contacts. The Spatial Data Transfer Standard was formalized as Federal Information Processing Standard 173 in 1992.

In 1994, President Clinton issued Executive Order 12906. It called for the development of a National Geospatial Clearinghouse through the FGDC.

## Privacy and Confidentiality

The Privacy Act of 1974 was designed to protect citizens from the internal misuse, and internal and external promulgation, of information about their person. Financial, health, employment, and legal attributes are example of Federal records that cannot be accessed by third parties. Individuals have the right to review records describing themselves for an accuracy check. Most states also have privacy laws, but their scope varies.

For environmental data, privacy is usually not a barrier. Cadastral land ownership records are primarily a state issue and are exempted from state privacy laws. Descriptions of the land, water, and air resources are not "personal." The one grey area involves resource transfer programs. On occasion, analysts studying non-point pollution are denied access to Department of Agriculture conservation, tillage, and chemical data because the records contain personal financial data.

Businesses often require a filter of confidentiality to protect trade secrets and business plans. Most environmental regulations requiring second party reporting have exceptions for requesting the designation "confidential information". This topic is discussed further in Chapters 7 and 8.

## Pricing

A potentially major hurdle (if not filter) affecting the dissemination of public environmental information is its cost. For Federal information, pricing policy has been well developed and standardized. OMB Circular No. A-130 establishes the basic principles that pricing should not create a barrier to access, that government has already incurred the cost of generating the information for public purposes, and that:

> the general standard for user charges for governmental information products, should be to recover no more than the cost of dissemination.

At the state and local levels, pricing is a highly controversial topic with many administrations looking at environmental information as a potential "profit center." Philosophies vary from profit, to system cost recovery, to cost of data reproduction. In addition to being guided by state, not Federal, policy, many agency information activities do not fall under the OMB assumption that system development costs have already been incurred. This is particularly true for agencies facing the transformation of paper maps to GIS. The Urban and Regional Information Systems Association tracks the development of state, provincial, and local pricing policies in its annual *Proceedings* series.

## STEWARDSHIP

### Sharing

The third major information policy framework is stewardship, which includes the topics of sharing, ownership, currency, and ultimately archiving or disposal.

In the Greenetowne scenario, it became obvious that a major obstacle to the development and wise use of environmental knowledge is the traditional Balkanization of data within narrow bureaucratic departments. Where the primary focus of information dissemination policy is the voluntary distribution of data between agencies and publics, the focus of information sharing policy is formal systematized intra- and inter-agency transfers. Basic variations are depicted in Figure 5.5.

Ad hoc data sharing is represented by the NEPA impact statement process. The draft EIS must be circulated for comment among other agencies which have a potential interest in or knowledge of a proposed project:

> Prior to making any detailed statement, the responsible Federal official shall consult with
> and obtain comments of any Federal agency which has jurisdiction by law or special
> expertise with respect to any environmental impact involved. (NEPA, Sec. 4332)

This requirement was a major breakthrough in the movement toward more comprehensive analyses. For example, the Corps of Engineers had for the first time to obtain ecological information from the Fish and Wildlife Service when assessing a proposed dam. However, in this approach, there is no standardization of content (such as classification) or format (such as GIS topology) for these comments. The process used, data analyzed, and feedback received varies from project to project.

For major improvements in efficiency and effectiveness, information sharing

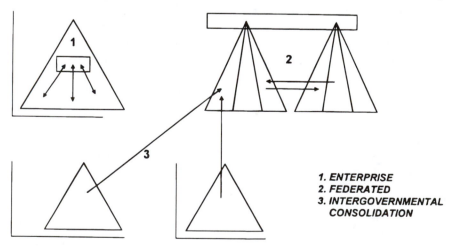

**Figure 5.5.  Information Sharing**

1. ENTERPRISE
2. FEDERATED
3. INTERGOVERNMENTAL
   CONSOLIDATION

between departments or agencies requires formalization. Within an organization "enterprise" data systems represent a unification of the logical design of the data structures and the networking of the various units. This represents a major policy challenge when multiple legacy systems exist. For example, the EPA's Toxic Release Inventory, (discussed further in Chapter 7) is a major success in providing unified chemical release information to communities and other external interest groups. Ironically, since most releases are both regulated and legal, much of the data which are generated by the massive annual survey process already existed within EPA in a series of independent, media-specific databases. The agency is slowly integrating these structures by standardizing the facility identification number key fields.

Policy has been promulgated to resolve this internal Balkanization. In 1990, the Agency issued an internal Order calling for a data standard for unique facility identifiers. This apparently was not fully implemented, in part because the legacy systems were designed to implement different facility definitions in the enabling regulatory legislation (Environmental Protection Agency, 1990). By 1995, in response to continued pressure for regulatory reform, the Agency proposed to consolidate annual facility reporting to a single form. A key to this proposed five-year transformation is that:

> each facility would be assigned a single identification number. Currently, each environmental program requires some form of facility identification, but those numbers are often not consistent identifiers among programs. ("Consolidated", 1995)

The above example deals with sharing related to a single data topic, permits. More ambitious approaches to comprehensive sharing include corporate and federated data models. An intra-agency corporate data model is based on a common data dictionary and common entity-relationships for all database information within an organization. In a networked environment, this would allow for "seamless" input, analysis, and reporting. From an information perspective, it would essentially remove the vertical and horizontal dataflow divisions of the traditional bureaucracy.

Due to the inertia of legacy systems and varying interpretations between units, such as regulatory and scientific, the corporate data model is more frequently a goal than a reality. The NY Department of Environmental Conservation has been designing its corporate data model using relational database concepts for over five years with no implementation in sight (Department of Environmental Conservation 1993). In contrast to New York's attempt to integrate existing relational databases, the Minnesota Department of Transportation has concluded that their corporate model requires the robustness of object-oriented tools. Only in this manner can they integrate the data needs of the multiple users and the physical systems hierarchy attributes that exist in the physical facilities and their contextual environments (Graphic Data, 1994).

A less ambitious but often more effective approach to comprehensive sharing is an inter-agency federation. This model is increasingly popular at the local government level for GIS implementation. The federation is formally established by the governmental units involved. Each agrees to use specified standardized data sets for transactional, tactical, and strategic activities. Generation and currency responsibility, along with "write-only" security access, for individual data layers is assigned among members. Policy also covers systems management and financing, including charge-backs.

A prime early example of the federated approach was the Alberta Land Related Information System. It combined four existing database and GIS sources of land-related data into a common data model that can be queried with a common language. A federated schema provides relationships between the data sources, while an external schema creates views for external users. The system is designed to serve agencies, corporations, and the public, and can be accessed remotely. In addition to user-friendly handling of queries, the system was designed to generate an estimate of cost prior to the processing of a request. (Goodman, 1994).

At the Federal level, a federated model is evolving more slowly. Executive Order 12906 (1994) states:

> each agency shall adopt internal procedures to insure that the agency accesses the Clearinghouse before it expends Federal Funds to collect or produce new geospatial data, to determine whether the information has already been collected by others, or whether cooperative efforts to obtain the data are possible.

Thus, the Clearinghouse is used by OMB as a filter. The continuing pressure for increased productivity is a major impetus for formalized data sharing. Some major progress has been made in basic data acquisition, such as jointly sponsored remote sensing. A fundamental obstacle exists when the data designs of different agencies are incompatible, as discussed in Chapter 9.

The most federated effort at the national level is the National Digital Geospatial Data Framework, called for in Section 5 of Executive Order 12906. The Framework, to be in place by the year 2000, is to contain geospatial data that are significant to a broad set of endusers. The proposed basic components include: Geodetic Control, Digital Orthoimagery, Elevations, Transportation, Hydrography, Governmental Units, and Casdastral (the Public Land Survey, and major public parcels). The Framework will support transactional updating. (Framework Working Group, 1994). Although the components of this federation are formalized, the use of the information is not required. It is expected that economics will serve this policy function.

### Ownership

Another key aspect of stewardship is information ownership. At the Federal level, the policy issue is simple: Federal data cannot be copyrighted. A complication arises when a Federal agency contracts with outside suppliers, a common occurrence in a period of outsourcing and downsizing. Here the case law varies (Doty, 1989).

In stark contrast, state and local governments may have the right to copyright. This power may be used to control content, by preventing unauthorized modifications, and to control resales, as discussed above under pricing. New York's Department of

Transportation has copyrighted its digital map series to help protect against unauthorized modifications. In addition to widely varying regional policies, the Internet is raising whole new areas of ownership issues.

## Currency

Environmental and social systems are in a continuous state of flux. Our ability to make
decisions is hampered by both obsolete data and the frequent lack of historical trend
information. Transactional information systems, such as the cadastre discussed in the
next chapter, are designed to be updated after each event. For some systems, like property deeds, the new record is appended to the historical record. For other transactions,
the new record replaces the previous one, drastically curtailing the ability of an analyst
to do detailed longitudinal research.

Although regulatory permit systems are transaction-based, regulatory jurisdictions based on "official maps," such as Greenetowne's wetland or floodplain programs,
are frequently frozen at the point in time of their creation. Even though the permits
often entail field visits, the changed conditions observed usually do not lead to a direct
map update. This significant currency issue is discussed further in Chapter 7.

The National Spatial Data Transfer Standard's metadata requirement to document
the lineage, date, and accuracy (at time of creation) of a data set only partially addresses the issue of currency (Federal Geographic Data Committee, 1994). Many of the initial datasets, which are digital versions of previously existing hard copies entered in the
system, designate these catagories "unknown." Because phenomena such as agricultural practices, suburbanization, and coastal erosion occur at different rates in different
locations, an independent analysis of change and data obsolescence is desirable. The
metadata standard is moot on this point. In the past, major environmental surveys have
been conducted "wall to wall" at periodic intervals. With the availability of advanced
information technologies, a shift toward continuous updating is possible. This transition will require a major change in policy perspectives.

## Archive and Disposal

A critical dimension in environmental knowledge building is the ability to learn from
past practices, and to be able to analyze longitudinal trends. Much of the nation's wetlands loss was the direct result of formal public agricultural and public health processes. The analysis of where and when these programs occurred is a vital link in the study
of historical habitat distributions. Thousands of industrial workers develop illnesses in
later life that are linked to exposures that occurred decades earlier. Establishing causal
links is essential to both the insurance claims processing and the development of regulatory programs. As agricultural practices become more sustainable by moving away
from the broadcast use of herbicides and pesticides, a question remains about the per-
sistence of previous applications in the environment.

In each of these examples, historical agency records are the key to knowledge building. OMB A-130 broadly covers each stage of the information life cycle. It calls upon all agencies to develop records schedules in conformance with National Archives and Records Administration and General Service Administration guidelines. The 1995 Paperwork Reduction Act reinforces the role of IRM in the entire life cycle (Poicher, 1996).

At the state level, New York's State Archives and Records Administration has been in the forefront of addressing these archival policy needs. The records of primary interest are those that are concerned with the decision processes leading to management's authorization of transactions" (NY State Archives, 1990, p. 2). All state and local agencies are to prepare retention and disposal schedules for all record series, such as environmental impact statements, land records, and permits. For each series, a determination is to be made regarding the minimum active retention period which meets administrative, legal, fiscal, and research needs. A process is established for systematically removing non-current records from physical and electronic files. These records are then placed in intermediate storage, typically within the agency. After an established period, the stored records are either destroyed or transferred to the State Archives for permanent storage. The latter decision entails a weighing of costs and potential information benefits.

The transition to electronic records has raised a series of major challenges to archival policy. These range from email to the rapid technological and physical obsolescence of storage formats. Environmental modelling and GISs pose significant problems. A digital dataset is typically created in a unique logical format that conforms with the software program used to analyze and display it. In a GIS, a thematic data layer, such as stream quality, is designed to be displayed as a map in association with a stored base map and stored legend and scale information. The software itself is designed to run under a particular operating system. When a decision is made to permanently archive the stream quality dataset because it will provide important historical benchmarks, what must be archived: The dataset, the associated data layers, the GIS program, the operating system?

In the preceding section on "currency", the argument was presented that, for well-informed decisions, policy should move from the historical episodic updating of environmental inventories to a continuous transaction-based updating. This approach raises important archival issues. A digital landuse map can be rapidly edited by a technician with a digitizer to reflect the change of a farm field to a subdivision. The edited polygon and its associated attribute record may constitute one millionth of the landuse dataset. What happened to the farm field? If 5% of the land use is updated every year the record set is never obsolete and therefore not subject to traditional archival policy. Clearly an approach is needed to store the localized historical records as they are superseeded.

In the discussion for the FGDC Framework, the traditional definition of "transaction", meaning a business event, was seen to be too limiting. "At a technical level, a transaction is any addition, deletion, or modification of a (GIS map) 'feature'" (Federal Geographic Data Committe, 1994, p. 4). No clear solution to this problem has emerged.

A three-stage approach would consist of first creating unique identifiers for adding data. Then an approach for tracking and holding changes would be needed, including feature-level metadata. Finally, a comprehensive approach for maintaining data history is needed. All of these are predicated on the emergence of object-oriented databases. If implemented, rather than viewing a static GIS map, we would be able to forward and reverse time and "play" the dataset like a video.

## SUMMARY

Information policy provides the context in which deep information can be realized. To date, information policy and environmental policy have been evolving in parallel. The responses to automation and the need for efficiencies are a first step. Environmental knowledge-building is a more complex process than simply data sharing. The development of integrated understandings of hierarchial environmental systems and how they change over time will require object data structures and corporate and federated IRM approaches.

Considerable policy progress has been made in the area of public pluralism. The rapid transformation of traditional access policies to the creation of the FGDC and the explosive growth of the Internet has fueled a major bootstrapping of the environmental information available to citizens, business, and interest groups.

The next four chapters will focus the policy lenses of this chapter on the major sets of legacy systems: Cadastral, regulatory, and science.

## REFERENCES

Burger, R. H. *Information Policy: A Framework for Evaluation and Policy Research*. Norwood, NJ: Ablex, 1993.

Committee on Government Operations, U.S. Congress. *A Citizen's Guide on Using the Freedom of Information Act and the Privacy Act of 1974 to Request Government Records*. New York: Science Associates, 1991.

"Consolidated Environmental Data Reporting Would Reduce Burden for Industry, EPA Says," *BNA Environment Reporter* (Feb. 17, 1995): 3,4).

Doty, P. "Federal Research and Development as Intellectual Property," in *U.S. Scientific and Technical Information Policies: Views and Perspectives*, edited by C. McClure and P. Hernon Norwood, NJ: Ablex, 1989, pp. 139–171.

Environmental Protection Agency. Order Classification No. 2180.3 Facility Identification Standard. (1990).

Environmental Equity Office, Environmental Protection Agency. *Fact Sheet: Environmental Equity Office*. Washington, D.C.: The Office, 1990.

Federal Geographic Data Committee. *Content Standards for Geospatial Metadata*. Washington, D.C.: The Committee, 1994.

Federal Geographic Data Committee. *NSDI Partnerships Working Group: Framework Session-Charleston*.Washington, D.C.: The Committee, 1994.

Foster, S. "Race(ial) Matters: The Quest for Environmental Justice," *Ecology Law Quarterly,* 20 (1995): 721–753.

Framework Working Group, Federal Geographic Data Committee. *Development of a National Geospatial Data Framework.* Washington, D.C.: The Committee, 1994.

Gerth, H. & C. W. Mills. *From Max Weber: Essays in Sociology.* New York: Oxford University Press, 1946.

Goodman, T. Alberta Land Related Information System: A Federated Data Base Case Study. URISA Proceedings. Washington, D.C.: Uraban and Regional Information Systems Assoc. (1994) 421:431.

Graphic Data Systems Corp. *Minnesota Department of Transportation Location Data Modelling Effort.* 1994.

Hernon, P. "Government Information: A Field in Need of Research and Analytical Studies," in *United States Government Information Policies,* edited by C. McClure, P. Hernon, and H. Relyea. Norwood, NJ: Ablex, 1989, pp. 3–24.

"Important Texas Watch Milestone Achieved," *Texas Watch* (Spring, 1994):1.

Jones, C. *An Introduction to the Study of Public Policy.* Monterey, CA: Brooks/Cole, 1984.

McClure, C., P. Hernon, H. Relyea. *U.S. Government Information Policies.* Norwood, NJ: Ablex, 1989.

NY Department of Environmental Conservation. *High Level Department-Wide Data Model— Draft.* Albany, NY: The Department, 1993.

NY Department of Environmental Conservation. *New York Citizen's Statewide Lake Association Program 1991 Annual Report.* Albany, NY: The Department, 1992.

NY State Archives and Records Administration. *Records Management and Internal Controls.* Albany, NY: State Education Department, 1990.

Office of Management and Budget. Circular A-16 Revised. (1990).

Office of Management and Budget. Proposed Revision of Circular A-130, Transmittal Memorandum No.1 (1992).

Office of Technology Assessment, U.S. Congress. *Informing the Nation: Federal Information Dissemination in an Electronic Era.* Washington, D.C.: GPO, 1988.

Poicher, D. "The Paperwork Reduction Act of 1995: A Second Chance for Information Resources Management," *Government Information Quarterly,* 13(1) (1996): 35–50.

Real, C., & R. Yoha. "California's New Seismic Mapping Act: Implications for Technology and Policy." *URISA Proceedings* (1991): 197–210.

Relyea, H. "Historical Development in Federal Information Policy," in *United States Government Information Policies,* edited by C. McClure, P. Hernon, and H. Relyea. Norwood, NJ: Ablex, 1989, 25–48.

Simon, H. A. *Administrative Behavior.* 3rd ed. New York: The Free Press, 1976.

# 6

Land Information Systems

## INTRODUCTION

*I*t can be seen from the perspective of the previous chapters that Greenetowne's endemic environmental crises stemmed in large part from inadequacies in systems information, knowledge development, and communication processes. Separate communities of actors and stakeholders based individual and organizational decisions on shallow, narrow, fragmented models of the environmental whole.

An early comprehensive analysis of this problem was articulated at the outset of the current environmental movement. Clawson and a group of resource economists were interested in analyzing the implications of the post-WWII suburbanization boom, one of the largest development periods in history. A major focus of concern was the rural-urban fringe, which was the target of highway, energy, residential, shopping center, office park, and other rapid construction activity. Questions included the effects on agriculture and natural areas, as well as the efficiency of "sprawl." They attempted to analyze the trends by reviewing data available from Federal and state agencies. Eventually they came to the realization that the existing information sources were inadequate both for their econometric modelling and for the private and public decision makers who were collectively shaping the rapid transition to a post-war America.

Clawson concluded that the information crisis he saw was the product of uncoordinated, narrowly focused data generation processes in the Federal and state agencies. As a response to the information gaps and Balkanization, Clawson postulated an integrated framework which would be necessary for comprehensively describing and modelling land use. The "layers" of this set include:

A. Natural qualities of the land, subsurface geology, surficial soils, water, vegetation and wildlife;
B. Improvements to the land, clearing, grading, drainage, infrastucture, and buildings;
C. Human activity on the land, agriculture, manufacturing, residential, and recreation;
D. Intensity of the activity, population density, agricultural practices, and manufacturing processes;
E. Land tenure, parcel ownership rights and distribution;

F.  Land jurisdictions, the multiplicity of governmental units and their associated taxes and regulations. (Clawson & Stewart, 1965)

These layers are depicted in Figure 6.1. Just having information in each category is an insufficient basis for analysis. Working at the threshold of the computer era and its complementary systems perspective, Clawson identified five dimensions critical to the integration of land information. First, spatial location entails the ability to correctly combine data from different scales and coordinate systems. Second, data standardization involves the development of common classification systems across jurisdictions.

Third, he called for a genius loci approach: that field data be highly annotated, and not limited to predefined codes. Here Clawson wrestled with the dilemma of standardization and the limited capacities of early data processing. What is necessary for large-scale aggregation frequently loses the rich situational uniqueness of a site. The fourth dimension, openness, involves the development of connections between land-related data and other emerging information systems. Clawson was interested in economic modelling so emphasized connections to other socio-economic data sets, such as the Census of Housing and Population.

Finally, the centrality of ownership: "Data from whatever source should be recorded separately for each parcel" (p. 5). Although he was primarily interested in the roles of Federal policies in shaping the management of natural resources, he was keenly aware that ultimately decisions were made by owners on a parcel by parcel basis. The central need for a parcel basis to integrate land information is the primary subject of this chapter.

Almost four decades later, now well into the second or third generation of computerization, the data fragmentation and Balkanization which confronted Clawson remains a central challenge. While GIS, databases, networking, and emerging Federal

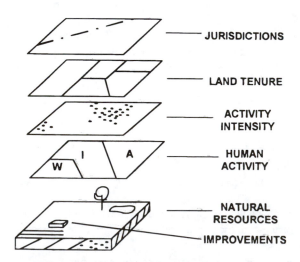

**Figure 6.1. Clawson's Land Information Layers**

and state policies have laid the foundation for some integration, few anticipated the explosion of concern and information relating to the environment, ecology, and pollution. Speaking of land use controls and environmental regulations in the early 1960s, he concluded: "By far the greater part of the total land area of the U.S. has no such limitations" (p. 25). Just the reverse would be true today.

Information Balkanization in the public and private sectors is not an ad hoc process. Policies and processes to generate, analyze, and disseminate information are the result of rational responses to institutionally legitimized needs. Because these institutions have specific foci and clientele, and have evolved at different periods, it is essential that we have some perspective on them in order to address the integrative topics which challenged Clawson.

## CADASTRAL HISTORY

In Browne County, the foundation of the socio-economic fabric is the pattern of land ownership. The white settlement of the area followed the surveying and platting of parcels. The plats, deeds, and mortgages constituted the first public information system, which has served continuously for over a century. As Clawson concluded, ownership parcels are the primary scale and thematic base map for organizing environmental information because the individual landowner is both the ultimate decision maker regarding activity type, intensity, and quality, and the legally responsible party in our highly regulated society.

The U.S. system of land records is deeply rooted in our British heritage. The development of the British system played a key role in that nation's transformation from a feudal to an industrial society.

The Anglo-Saxon period in England was characterized by small villages surrounded by two types of land patterns. Large communal open fields for grazing and strips of row crops typified the Midlands, while in other regions small agricultural fields bounded by hedgerows or stonewalls were the result of agricultural expansion into forested, fen, moorland, and marsh environments. Within about a mile radius from a village, one would encounter, "the edge of the wild" (Hoskins 1970,p. 82). Growth in the economic base came from both increased productivity in established areas and expansion into uninhabited territories. Major contractions occurred during periods of political turmoil, plague, and climate change.

From an environmental information perspective, the feudal system was particularly "dark." Except for primitive maps, there was no clear picture of the environmental resource. Parcel level data were nonexistent. The freehold system and its resulting crazyquilt of tenure classes and taxation responsibilities led to numerous ad hoc approaches to enumeration of fields, livestock, produce, and serfs. There was no organized science based on field-collected data. "Advances" made in agriculture and development, such as draining of swamps, were purely vernacular.

The feudal system began its decline with the signing of the Magna Carta in 1215.

Nobles gained more control of the land and the system of common law began to coalesce. The monarch retained the right to military service and to tax the land. Labor flexibility was introduced as some serfs were allowed to become peasants who could move, paying rent or tribute in exchange for occupying and working a plot. Of the four original freehold types, "socage" persisted and evolved into the modern construct of "estate." Citizens, and later corporations, obtained the right to "own" (acquire, use, sell) parcels of land.

British capitalism emerged forcefully in the form of the woolen industry which utilized England's productive grazing lands and robust shipping capabilities to generate a leading position in a growing world market. Capitalism requires fluid basic resources, including labor, land, and money. The village, commons, and strip crop patterns which were ingrained in the feudal landscape, and had been divided into smaller and smaller patterns due to generations of inheritance, were inefficient for large-scale sheep farming. The Enclosure Acts of the 1700s fundamentally broke the old system and facilitated the assembly of large single-ownership grazing enclosures.

Between the 16th and 18th Centuries an economic paradigm shift took place. The land resource was recast as a critical commercial input. The medieval cadastre of inheritance, royal charters, and field-placed monuments was incapable of supporting the land transactions that capitalism demanded. For example, a traditional conveyance had entailed a "livery in deed," a paganish ritual in which both parties walked the boundaries, with rites at each major corner point (note: it is interesting to compare this in situ process with today's "innocent purchaser" hazardous waste site audits discussed in Chapter 8).

The institutional transformation of land into a private market commodity carried with it all the potential abuses associated with unfettered commerce, which was often conducted at locations far from the actual site in question. Much of the difficulty was informational, either joint ignorance of the resource, or misrepresentation of the resource to the buyer. To control these free market failures, which could undermine the new economic system, two major reforms were enacted. The Statute of Frauds (1677) required that all land contracts must be in writing and signed by both parties. By itself, this Act failed to curb the practice of multiple conflicting sales of the same parcel which often didn't come to light for years after the transactions. The Fraud Act was therefore supplemented by the Statute of Enrollments which rendered written conveyances of real property void unless recorded within six months at the designated local or regional court (Brown, 1962).

These information policies had a profound effect by creating a public information environment that was in concert with the emerging private capitalism. The records, primarily deeds and plats, became part of an information system of input, storage, and retrieval. As new conveyances took place, these records were added chronologically to the public holdings. Each new record was linked by citation to its predecessor, creating a "chain of title." To facilitate record access, simple indices of buyers and sellers were developed. Private buyers, sellers, lenders, and insurers could readily access the decentralized storage sites and analyze the documents prior to making a transaction. In addition to promoting commerce, the system served the further function of continuously updating property tax rolls, the major source of revenue for local governments. The power and simplicity of this information system has endured, as can be witnessed in hundreds of local courthouses daily throughout the U.S. The set of public records which deal with land ownership is called the "cadastre."

If one examines the contents of a deed it becomes obvious why a chain of title is necessary. In stark contrast to Clawson's ideal comprehensive set of descriptors necessary to fully describe the land, a deed description is typically limited to brief enumeration of political jurisdiction, current owner, and property location, such as survey measurements or naming of bounding parcels (Pedowitz, 1984). There is no description of the natural resources, or land "improvements," or type and intensity of human activities. While Clawson realized it may be impossible to describe fully all the detailed physical characteristics of a land parcel, in a basic deed the subject of the environment is intentionally ignored.

In place of physical descriptors is the purely capitalistic construct of divisible ownership rights. If land is a commercial commodity, it can be used in many ways by many parties over time. The answer to how to efficiently describe this is embodied in the legal concept of "fee simple." Fee simple ownership connotes complete ownership of a parcel, the soil and minerals below, the air space immediately above, all entities attached to land such as trees and buildings, the rights to full use of the land, and the ability to convey some or all of these to others. A standard deed description begins: "All that certain plot, piece or parcel of land, with the buildings and improvements thereon situate lying" followed by the location description, followed by the exceptions and reservations (p. 247).

The fee simple represents a "complete bundle of rights" which can be divided. For example, a parcel may contain easements dedicated to a public highway and a utility company. The mineral rights for underlying coal may have been sold off, and a contract signed for long-term timber harvesting. The property may contain access and use rights to a lake on an adjoining parcel. Over the years, a parcel's original fee simple bundle may be highly split through a large set of individual transactions. A new deed has to list explicitly only any new exceptions or reservations. All prior reductions or expansions in fee are contained in reference to the chain of title by simply citing the previous deed. When a party buys land rights, what are they buying? To legally answer this question, the cadastral system facilitates a title search, supplemented by a cross-indexed mortgage and lien search to identify outstanding debts in which the property rights have been used as collateral.

The construct of fee simple conveniently obviates the need to describe the environmental resources in order to conduct a property transaction. Are the orchards maintained?; is the soil eroded?; are the buildings sound?; is the well polluted?; is the residence in a flood zone? A deed, by design, ignores such questions. In classic eighteenth-century free market form, the environmental corollary to the institution of a cadastral information system is caveat emptor.

Deed and Mortgage systems are controlled by state and local law. Jurisdictions all over the U.S. are currently modernizing these systems. A first approach is to transform the indices to an online database. A second stage is the full-scale scanning of text. A third stage involves networked remote access. These modernization steps greatly enhance transactional efficiency, but by themselves do little to build environmental systems knowledge.

## REAL PROPERTY TAX RECORDS

In contrast to the information contained in deeds, property tax records represent a rich but skewed set of environmental information. Deed information is almost entirely second party, a standardized record of a private contract. Real property tax assessments consist primarily of first party data collected on site by the assessor. The primary objectives of property assessment are to maximize the tax yield while maintaining equality and equity. Since real property includes both the natural resource base and fixed physical improvements, a wide range of attribute data may be generated for each parcel.

Property tax data may include portions of most of Clawson's information layers. Most local services are supported primarily or in part by the property tax. A landowner in Greenetowne may simultaneously be in a county, a town, a school district, a water district, a sewer district, and a park district, each with different spatial jurisdictional bounds and each with authorization to tax by applying a rate to the property's assessed valuation.

A typical taxation record for a parcel will contain: a unique parcel number; owner name and address; list of tax jurisdictions; area and critical dimensions such as length of road or water frontage; land use classification; and detailed data regarding improvements. For example, a residential parcel will include a description of age, style, size, materials, upkeep, and interior configuration of the home. Similar data are collected for commercial and industrial properties.

Classification and its intrinsic policy issues were introduced in Chapter 4. It is a controversial topic when it is the basis for significant decisions. For property taxation, two classification policy issues arise. First, many properties have more than one use. A farm may include a residence, field crops, a woodlot, and a roadside commercial stand. A suburban building may have a commercial parking lot, offices on the first floor, and apartments above. Although the manual records may have detailed written descriptions of these attributes, it is common for computerized records to utilize a single primary classification, or a dual primary and secondary classification.

The second major classification policy issue is time frame. Should a property be classified according to current use or potential use? Here the bias of the system is quite clear. The entire purpose of generating the descriptive information is to allow an estimate of the "fair market value" of a parcel. The way this is done is to statistically compare its attributes with those of similar properties for which recent sale prices are available. In a residential area, the current and future land uses are often the same. For farmlands in a metropolitan fringe, however, the "highest and best" use (reflected in market price) may be represented by intensive development. Thus the tax classification may not only not represent current reality, but its resulting increase in carrying costs may accelerate the development process in a kind of self-fulfilling prophecy.

These classification issues may severely limit the use of tax data in environmental analyses. For example, watershed models are used to predict the quantity and quality of storm runoff. They require spatially distributed land use data. While tax data may be readily available, the land uses represented by primary classifications and/or highest and best use assigned to the entire parcel area may not accurately describe the physical properties of the watershed.

Fair market value is the basis of property assessment, but, due to issues of equity and politics, all properties are not assessed at this value. Throughout the history of this country, full exemptions has been granted for public and religious properties. In this century, as property taxes have increased, an extensive list of partial exemptions have been legislated by the states. Some of these are statewide mandates while others are permissive delegations to the local taxing district. Exemptions may deal directly with environmental factors. Examples include wetlands, wildlife sanctuaries, managed forests, and prime agricultural districts.

Can an environmental analysis effectively use information from real property exemption files? The answer to this question is not straightforward. Consider a hypothetical Greenetowne example. A watershed modeler wants to predict water quality in fishing streams. Assume the state has a permissive partial exemption to encourage the use of Best Management Practices (BMPs, such as erosion control) in harvesting from private woodlots. The modeler accesses the tax database and identifies participating parcels and their associated acreage in the watershed. Two major data obstacles confront the modeler. Many woodlot owners may have chosen not to participate. Acreage may not be in the tax files, depending on the classification used (see above). Some of the owners may be using BMPs voluntarily; others may not.

Of the participating properties, the primary source of the information frequently is one time (exemption application), second party self-reporting. The performance-based tax exemption shares a basic flaw with environmental regulations discussed in the next chapter. The lack of periodic first party monitoring often renders cadastral environmental data of dubious quality. This problem not only affects our modeler. More significantly, it means that policy makers often have inadequate program effectiveness information.

## PROPERTY AND TAX MAPS

As the U.S. was settled, a fundamental challenge to the governments was the efficient transfer of property from private to private, and public to private, sectors. Delineation of property boundaries is critical to effective transfers. Generalizing, property ownership mapping falls into two basic categories: metes and bounds, and the Public Land Survey (PLS). Most of the east coast, Texas and Hawaii are "metes and bounds states," while the rest of the states are in the PLS.

A metes and bounds survey typically involves measurement of the angles and distances around the perimeter of a parcel (metes), and the identification of adjacent landowners (bounds). Significantly, a metes and bounds survey typically "describes a parcel of land as though it were an island standing on its own" without reference to some comprehensive coordinate system (Brown, 1989, p. 5-3).

A metes and bounds-based deed usually doesn't include a map. Maps are, however, frequently used as efficient means of describing subdivisions. A formally filed subdivision plat will include metes and bounds of the original parcel and metes of each new parcel, with the new parcels being coded by lot numbers. The deed location description can then be condensed to simply naming the filed plat and citing the lot number. The entire deed recording system is focused on the individual parcel, with no consideration given to the next broader system scale.

Over centuries, the deed transfer system has generated a spatially uncoordinated verbal and pictorial maze of millions of metes descriptions, bounds descriptions, and subdivision plats. Not only do the deeds not lend themselves to an aggregate view of a locale, they frequently having conflicting descriptions of border locations. Early on, many municipalities realized the need for a comprehensive parcel mapping approach. The development of Manhattan in the 19th Century was predicated on the systematic mapping of blocks and lots with the accompanying development of infrastructure. In the past decade, numerous local governments have undertaken land data modernization projects which have included both large scale comprehensive base mapping, such as orthophotos, and an analysis of deeds.

In 30 states, covering almost 80% of the continental U.S. land area, the Federal government utilized the Public Land Survey (PLS) to manage its holdings and facilitate the transfer of government property to states and private parties. The PLS was established in 1785. Drawing on the surveying experience of private land developers in Maryland, Pennsylvania, and western New York, the survey design reflects a unique combination of economic and philosophical factors.

Utilizing Jeffersonian modernism, the basic scale of the program was to "lay out mile square parcels over all the federal lands…" (Speight & Abrams, 1989, p.6-2). As was seen in Greenetowne, the regularity of the geometry was the ultimate commoditization of the spatially complex environmental resource, which intentionally ignored both natural features and the existence of the land's native inhabitants. The only exceptions were the "meanders" allowed for navigable-scale waters. The one-mile units constituted "sections" of a "township" which was composed of a 6x6 section square. The sections were frequently divided into quarter sections of approximately 160 acres, and these could be subsequently halved and quartered. The use of sections and subsections were directly related to the state of agricultural production. One-family farms using horse or oxen power equated to the quarter section which was the primary parcel of the 1862 Homestead Act.

Where the land was settled, many of these surveying townships later became legitimized as municipal townships. The next largest scale unit was the county, which served to provide local distribution of state functions, such as courts and taxes. The location of the county boundaries and the county seat activities were the products of territorial or state legislators. These often utilized but didn't necessarily follow the PLS hierarchial framework.

Even at the township scale, the earth's curvature manifests itself. Geographically, the northern boundary line of a "square" township is 20 to 50 ft. shorter than the southern bound. Reflecting the accuracy of available instruments, the practicalities of field time, and the primary agricultural land use, the standard surveying practice was to approximately maintain the one-mile squares and to periodically make adjustments. These adjustments were made at the perimeters of 16 town (4x4) blocks. The southern four towns in a block had one mile wide southern borders.

Thus, in the PLS states, a comprehensive mapping framework existed from the outset. At one time the Federal government had title to approximately 80% of the 2.3 billion acres in the 50 states. It currently owns about 30%, 730 million acres (p. 6). The transfers of parcels to the states and to private owners took a variety of forms, sometimes in large quantities such as the subsidies granted to induce railroad construction or subsidies for the development of state landgrant colleges, and sometimes in small parcels such as homesteads.

Subsequent owners could divide and sell their parcels just as owners in the eastern states had done since the colonial period. Over time, the simple PLS grid was overlayed by a collage of deed and plat metes and bounds descriptions. Farmers and developers responded to local natural features and market demands, frequently diverging from Jefferson's idealized rectilinear geometry.

Even in the nationally held lands, the fee simple title was continuously modified. National parks and wilderness areas were legislated. Thousands of mineral, grazing, logging, and utility corridor leases and agreements were authorized. Each of these transactions involved descriptions and maps that were related to but differed from the PLS grid. By the 1960s the PLS states and the Federal government were in agreement with the non-PLS states that a comprehensive approach to parcel mapping was needed in order to effectively visualize and manage environmental resources, to meet the growing needs for public services and to provide the necessary information required to support the business activities of the private sector.

> Every knowledgeable person having contact with our system of land records knows that the strain upon it is growing in geometric proportions every year. (Basye, in Moyer & Fisher, 1973, p.1)

## PARCEL IDENTIFICATION NUMBERS

From a systems perspective, the traditional deed and mortgage cadastre was both internally atomized and externally isolated. Parcels records could not be spatially aggregated to provide a systems view of contiguous parcels in a management region, such as a watershed or service district; nor could the other non-cadastral land use attributes like subsurface geology or vegetation, identified by Clawson, be readily integrated with the parcel ownership data.

The late 1960S and early 1970s were both a period of cadastral information crisis and the advent of large-scale public and commercial computing. The American Bar Association took the lead by organizing a series of conferences starting in 1966. Representation ranged from Federal land and development agencies to state and local governments, assessors, banks, lending institutions, insurers, and academics. After considerable study,

> It became evident that the chief obstacle to records improvement was the lack of a common system of land parcel identifiers and that any universal system must be compatible for application to other land related records (land use, ecology, etc.) as well as to land title records. (p. xvii)

The final conference in 1972 focussed on "Compatible Land Identifiers." The sessions built upon a number of invited studies which had been prepared from the analyses of the previous meetings. A variety of approaches were reviewed. Street addresses, Grantee–Grantor Indexes and other non-coordinate systems were rejected for a host of analytical and implementation reasons. A review of seven PLS states revealed considerable variation in the use of county, township, and range designators, and the need to develop identifiers for subdivisions, blocks, and parcels below the quarter section.

With the focus shifted to coordinate-based descriptors, the central policy issue was the monumental cost and time of fully resurveying every parcel to get a complete and accurate polygon outline versus the cost-effectiveness of a more simplified descriptor. The recommended basic universal parcel identification number consists of the municipal FIPS code (Federal Information Processing Standard 6-2, 1973), and the State Plane Coordinates for the visual centroid of the parcel. A mathematical centroid is the position at which a cutout of the parcel shape would balance on the point of a pin. To compute it, a full numerical description of the perimeter is required. There are many situations, such a "c"-shaped lot, where the mathematical centroid actually falls outside the parcel bound. In contrast, the visual centroid can be quickly placed by the digitizer viewing a parcel map and simple coding rules can consistently keep the point in the parcel. The Conference recommendations were simple and comprehensive. Each parcel requires a unique Parcel Identification Number (PIN), consisting of county FIPS code and state plane coordinate visual centroid, which should be included in all cadastral records, including deeds, mortgages, and property tax systems. Although many case studies were utilized in the Conference sessions, little formal attention was given to implementation of the recommendations. New York State was used as a new model program cited by the Conference. The recommended standard is illustrated in Figure 6.2. Its pre- and post-conference experience provides many useful insights.

In the 1960s, New York State underwent explosive suburbanization. To efficiently provide for this growth, state policy encouraged the formation of large consolidated districts for the provision of schools, water, and sewers. These services were primarily

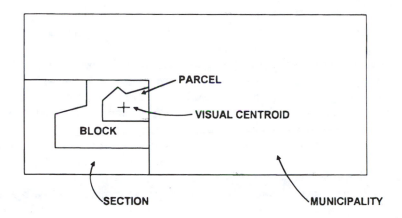

PIN = STATE, COUNTY, TOWN, SECTION, BLOCK, PARCEL, CENTROID X, CENTROID Y

**Figure 6.2. Parcel Identification Numbers**

financed through the property tax, which traditionally had been administered on a local municipality basis. A crisis, both political and legal, soon emerged due to differences in taxation practices within jurisdictions compromising the newly constituted consolidated districts.

The solution was to develop a standard statewide system of assessment. In 1970, the state legislature passed a comprehensive bill, "Preparation and Maintenance of Tax Maps for Real Property Assessment and Taxation Administration" [, 1971, amended #115]. Each county was charged with the responsibility to prepare a new set of standard tax maps. The maps were to be based on aerial photography, using reference Coordinates from the State Plane coordinate system. Four scales were available, 1" = 400', 200', 100', and 50', depending on the density of development. This scale range, "micro planning" to "general design," was selected so that each parcel could be efficiently depicted and annotated. A standard sheet size was specified, and sheets were named "sections." Within section maps, blocks were delineated utilizing physical and cultural boundaries. The parcel was the basic mapping unit. Parcels were assigned a unique PIN, consisting of municipal code and section-block-lot number. For each parcel, a visual centroid in State Plane Coordinates was generated. Extensive deed searches were utilized in assisting the mapping process.

The mapping was the foundation of a comprehensive computer-assisted mass appraisal system. A Board of Equalization and Assessment was established to annually generate equalization rates for each taxing district to ensure equality of burden. Local assessors were provided with standard classifications and reporting forms. Each subdivision resulted in the generation of new PINs. Each sale or building permit was recorded in a standard format. Counties were mandated to develop and maintain a standard database on tax parcel assessment information, and an updated copy was sent each year to the state Board. The Board would then use statistical sampling, including independent field audits, to ascertain and manage equalization equivalence between assessing jurisdictions (Kitchen, 1987).

This comprehensive property tax information system has been highly successful in meeting the legislature's objectives. The development of a statewide GIS that includes municipal and tax district boundaries has enabled the Board's staff to rapidly analyze complex spatial problems by conducting "point in polygon" models in which the outline of a area in question is overlaid on a mapping of parcel centroids, and the attributes of identified parcels are then extracted from the tax database. This centralized system was developed to solve a particular financial problem and does not meet the objectives of a multi-purpose LIS.

Four major criticisms from the real estate and finance communities have been that:

1. Cadastral deed, mortage, and map information are not digital;
2. Required environmental information from other state agencies is not digital;
3. The data sets in (1) and (2) are not linked with the spatial tax parcel information system; and
4. This information is not available remotely for access and Electronic Data Interchange transactions.

In 1993, a Governor's Task Force on Filing and Recording developed the following future scenario:

An attorney in Manhattan working for clients in San Francisco is asked to study alternative plant locations on Long Island for a rapid investment decision. From her desk she connects to the "New York State Land Information System" and, using a mouse, zooms into a series of maps to the local area of interest. By keying in parcel size and other criteria, a list of potentially suitable parcels and their PINs is generated. The attorney then requests that these parcels be checked against databases in the state's Environment, Health, and Agriculture Departments for hazardous waste, wetlands, radon, and agricultural chemicals. For those parcels that pass this screening, connections are then made to the County Clerk and County Real Property Offices and information assembled on ownership, mortgages, taxes, and zoning. Within a couple of hours, the analysis with the attorney's evaluation is e-mailed to California (NY Governor's, 1993).

Although the 1971 law had mandated the use of PINs for tax records, it said nothing about their use in deeds, mortgages, or subdivision plats, or environmental records. The tax records did include the deed book and page reference that was generated at the time of map preparation. The Task Force strongly recommended the universal adoption of "uniform parcel identification" as a critical element in facilitating the GIS integration of LIS:

The land records system in New York state is making a difficult transition from its seventeenth century paper based roots to a twenty first century electronic future. (p. 30)

The Task Force recommended the creation of a new state coordinating body and extensive revisions to existing legislation. Although the above scenario hinged on the inclusion of three major non-cadastre environmentally focussed departments, the report made no specific recommendations regarding development of an information federation. To date, there has been no substantive followup to the Task Force's recommendations.

## LAND INFORMATION SYSTEMS

From the perspective of sustainability, land ownership is a critical but woefully incomplete description of the environment. Whereas deeds intentionally omit environmental attributes, tax records view elements such as buildings, fertile soils, and water frontage from the highly skewed perspectives of market value and myriad legislated revenue exceptions, such as wildlife preserves and prime farmland. From Clawson to the ABA Conference, the rapid increase in computer hardware and software advances fueled interest in the promise of a multi-purpose cadastre which integrated land records, environmental, and socio-economic information. The advent of satellite remote sensing in the 1970s, and U.S. Bureau of the Census' vector GIS file developments for the 1970 census, added to the momentum.

In 1980, the Federal Committee on Geodesy of the National Research Council issued its report, *The Need For a Multipurpose Cadastre*. After concluding that such a system is necessary for the cost-effective delivery of public services, the Committee identified the four basic components of a multi-purpose cadastre:

1. Geodetic coordinate reference framework;
2. Current, accurate large scale base maps;
3. A parcel boundary overlay, with each parcel assigned a unique PIN; and
4. A series of registers or files which link other related land records through the use of PINs.(Committee on Geodesy, 1980)

Figure 6.3 depicts the integrative form of the system, and the role of PINs.

The Committee, by definition, had a bias toward accurate surveying. Items 1,2, and the first part of 3 above have major policy implications for the Federal and state governments in providing uniform, high-quality mapping control points and standards. The need for new Federal legislation was identified. OMB was called on to designate a lead Federal agency to oversee the development of multi-purpose cadastre systems. State governments were asked to play the lead role in modernizing tax mapping, parcel identification, and other cadastral policies. In a follow-up report, *Procedures and Standards for a Multipurpose Cadastre*, the counties are called upon to provide the lead in the actual integration of the "other related land records," item 4 above (Committee on Geodesy, 1983). This reflects both their traditional role as the keeper of cadastral records, and the computer technology of the period, when counties were

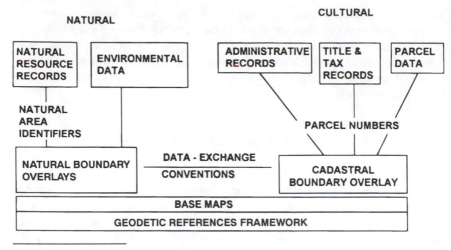

NATURAL                                      CULTURAL

a based on Committee on Geodesy, 1983, p. 16

**Figure 6.3.  Components of a Multipurpose Cadastre[a]**

developing centralized data processing functions, and operational high-speed networks did not exist. It ignores the fact that the vast majority of environmental regulation records are Federal and state policies.

The essence of a multi-purpose system is that transactional data are to be continuously generated and shared in a common federated or enterprise computing environment. Both reports allude to the endemic institutional barriers to such an environment, but neither study substantively addressed this central policy issue.

The term cadastre is both quite old, and unfamiliar to most of the public and elected officials. With the popularity of the acronym GIS to broadly include almost any computerized spatial information system, it is only natural for professionals working with ownership-related data to seek a clearer, more current term. Land Information Systems, LIS, has in recent years been replacing cadastre. The Federal Geodetic Control Subcommittee (FGCS), which was charged with helping implement the recommendations of the NRC Committee, uses the term Multipurpose LIS (MPLIS) to emphasize that the cadastral parcel should be the common bridge for integrating natural resource and cultural data sets, particularly at the level of local government. The FGCS has produced *Multipurpose Land Information Systems-The Guidebook* (Brown & Moyer, 1989) to facilitate the implementation of county and municipal LIS.

Many local governments are developing GIS and LIS approaches. These are frequently highly independent undertakings. Wisconsin has been a leader in generating a statewide "top down" framework which promotes "bottom up" county by county implementation. This formal approach was the result of over a decade of state-university-county working relationships and demonstration projects (Niemann, Jr., 1990).

Unlike New York, which entered the LIS arena with a narrow, single-purpose, single-agency property tax focus, Wisconsin's system has evolved from a broad multi-purpose tradition. Starting with a series of county scale demonstration projects in the late 1970s and early 1980s, a group of Federal, state, and local agencies under the coordination of the University of Wisconsin developed working LISs, which integrated cadastral, natural resource, and agricultural data. Pilot projects focused on combined information technology and policy to address critical topics such as watershed water quality. These successful and highly visible products provided the impetus for a Governor's executive order creating the Wisconsin Land Records Committee in 1984. In 1989, the state legislature established a Land Information Board to implement the Land Information Program.

Functionally, the Governor appoints a Land Information Board which is associated with but acts independently of the State Department of Administration. The Board inventories information systems, recommends legislation, develops standards, approves county land record modernization plans, and operates the local government LIS aid program. The Office of Land Information is the small central professional staff to the Board. Counties may participate in the program by formally creating a Land Information Office; developing a plan that includes a minimum of five elements (geographic reference, parcels, wetlands, soils, and zoning); and providing annual progress reports. The public state and county organizations are complemented by a non-governmental organization, the Land Information Association, which includes individuals, firms, and agencies. Its agenda includes continuing education and dialog among all affected actors and stakeholders, and lobbying for support and legislation (Moyer & Niemann, Jr., 1989).

The Wisconsin approach is clearly more comprehensive than New York's. Their visual schematic model is essentially the same as Figure 6.1. Even here however, large components of the existing environmental data sets are not being integrated. The primary focus has been modernizing of county geodetic base mapping control and cadastral data at the county level. With the exceptions of soils and wetlands, Federal and state agencies involved in the regulation of air, water, and land pollution have not to date formally recognized this transformation by adopting policies which require PINs in their regulatory permit records. This omission is consistent with the lack of direct integration in the Federal policy model depicted in Figure 6.3. Here the environmental records are connected to parcel records via a GIS spatial overlay, not by means of a PIN-based relational database key field. The (unimplemented) New York proposal is a considerably more robust approach to deep environmental knowledge building.

The vast majority of LIS activity in the U.S. is occurring at the local government level, typically without the comprehensive Wisconsin-type state framework. These approaches were discussed briefly in Chapter 5 under data sharing. Cumulatively, the experiences of these grassroots efforts are generating some pragmatic implementation guidelines, but serious challenges have arisen, often linked to high startup costs and the resistence to the fundamental organizational changes upon which a federated data model is predicated (Onsrud & Rushton, 1995).

## SUMMARY

While a GIS with associated predictive models may adequately describe the physical conditions of the environment, a multi-purpose LIS is the only comprehensive, practical way to inform the entire continuum of decision makers about the land's environmental condition and their responsibilities for stewardship. In addition to being the only appropriate scale in which to engage land responsibility issues, the ownership component of an LIS can track transactions. This internal capability of continuous updating is a key to monitoring change over time.

For deep information, a parcel-based data structure is essential. It provides the critical link to owners, investors, and insurers which bridges the centuries-old chasm between property ownership and the physical environment that ushered in the era of industrialization and environmental externalities. A fundamental challenge to deep information is the continuing Federal and state policy vacuum regarding the integration of parcel numbering and regulatory information. This topic is dealt with in more detail in the next chapter.

## REFERENCES

Brown, C. *Boundary Control and Legal Principles*. New York: Wiley, 1962.

Brown,C., W. Robillard, & D.Wilson. *Boundary Control and Legal Principles*. 3rd ed. New York: Wiley, 1986.

Brown, P. Property Boundaries [Supplement]. In *Multipurpose Land Information System: The Guidebook*. edited by P. Brown and D. Moyer. Washington, D.C: Federal Geodetic Control Committee, 1989.

Brown, P. and D. Moyer, ed. *Multipurpose Land Information Systems: The Guidebook*. Washington, D.C.: Federal Geodetic Control Committee.

Clawson, M. & C. Stewart. *Land Use Information: A Critical Survey of U.S. Statistics Including Possibilities for Greater Uniformity*. Baltimore, MO: Resources for the Future, 1965.

Committee on Geodesy, National Research Council. *Need for a Multipurpose Cadastre*. Washington, D.C.: National Academy Press, 1980.

Committee on Geodesy, National Research Council. *Procedures and Standards for a Multipurpose Cadastre*. Washington, D.C.: National Academy Press, 1983.

Hoskins, W. *The Making of the English Landscape*. Middlesex: Penguin, 1990.

Kitchen, R. "Geographic Reference and Spatial Analysis of Assessment Information," in *Proceedings: International Conference on Assessment Administration*. I.C.A.A. New Orleans, 1987.

Moyer, D., & K. Fisher. *Land Parcel Identifiers for Information Systems*. Chicago: American Bar Foundation, 1973.

Moyer, D., & B. Niemann Jr. "Institutional Arrangements and Economic Impacts," in *Multipurpose Land Information Systems: The Guidebook*, edited by P. Brown and D. Moyer. Washington, D.C.: Federal Geodetic Control Committee, 1989.

NY Governor's Task Force on Filing and Recording. *Public Records: Vision 2000- Land Information*. Albany, NY: N.Y.S. Forum for Information Resource Manangement, 1993.

Newton, P., et al., ed. *Networking Spatial Information Systems*. New York: Belhaven, 1992.

Niemann, Jr., B. et.al. "Land Information Systems Modernization in Wisconsin: Government—University—Professional Interactions." *Government Information Quarterly,* 7 (3) (1990): 269–284.

Onsrud, H., & G. Rushton, ed. *Sharing Geographic Information*. New Brunswick, NJ: Rutgers Center for Urban Policy Research, 1995.

Pedowitz, J., ed. *Real Estate Titles*. Albany, NY: N.Y.S. Bar Association, 1984.

Speight, G., & J. Abrams. "The Public Land Survey System," in *Multipurpose Land Information Systems: The Guidebook*, edited by P. Brown and D. Moyer. Washington, D.C.: Federal Geodetic Control Committee, 1989.

# 7

# Environmental Regulation

## INTRODUCTION

Our most extensive source of environmental information is contained in the numerous Federal, state, and local regulatory systems which document the public control of human activities. This source is highly fragmented due to the multiple jurisdictions involved and the legislative and administrative tendency of regulations to focus on narrow problems.

Environmental regulations are also the most highly contentious source of environmental information. Both landowners and industry have systematically opposed regulations, even those that have been demonstrated over time to efficiently improve health, sustainability, and property values. The opposition ranges from "paperwork reduction," the simplification of self-reporting, to de-regulation. Environmentalists have frequently criticized the limited scope of regulations, and the lack of meaningful enforcement. Scientists often conclude that regulations are a knee-jerk reaction to a crisis, and are not built on a sound foundation of scientific understanding. Minorities increasingly voice the opinion that regulations regarding allowable exposures, facility siting, monitoring, and enforcement are inherently biased. Finally, bureaucrats complain that laws are often either too specific or too fuzzy, that implementation resources are endemicly inadequate, and that the frequent resort of actors and stakeholders to litigation overwhelms a stressed administrative system.

The environmental regulation of human activities is an ancient societal practice. Agrarian cultures controlled irrigation flows. The Romans regulated road design and solar access. Societies have banned or zoned health and safety nuisances for millennia. Modern regulations are seen as a public solution to private sector market failure, such as industrial discharges, which traditionally was characterized as an economic "externality." Ideally, environmental controls involve a sequence of information flows, as depicted in Figure 7.1. Legitimization includes the transformation from problem perception to formal law and rule making as discussed in Chapter 5. This is followed by an initialization stage. Often, spatial and substantive jurisdictions need to be articulated. The administration attempts to inform the regulated community of the problem, the jurisdictions, and the procedures of the regulation, such as a permit process. The implementation stage includes permit processing, monitoring, and enforcement. Systematic

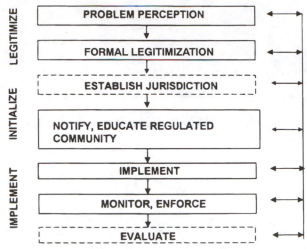

Figure 7.1. Information Phases of Environmental Regulation

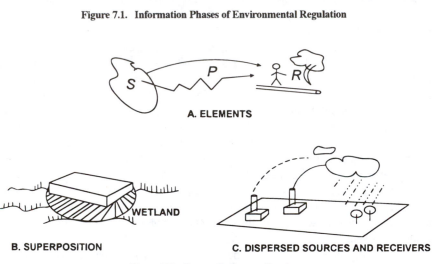

Figure 7.2. Source–Pathways–Receiver

effectiveness evaluation involves feedback information from each program step as a basis for continuing improvement.

The framing of the environmental "problem" has traditionally taken the form of a simple, mechanistic cause-effect linkage of source-pathways-receiver as shown in Figure 7.2a. The source is a generator of energy and matter, such as a construction project or industrial emissions. The pathways are air, water, surface, and subsurface transmissions (and modifications) of the energy and matter such as a floodway, or disease-carrying animal and insect vectors. Finally, the receiver of interest can be a physical system such as storm drains, an ecological system like a wetland, or human systems, particularly individual health. The rationale for a regulation is to prevent or reduce harmful stresses on a valued receiver. Operationalizing what level is harmful (or acceptable) is a major focus of debate involving both complex systems issues and the pragmatic politics of costs, distributions, and feasibility. The vast majority of environmental regulations involve the formalization of either nominal (e.g. endangered species, wetlands) or ratio scale numerical (e.g. traffic volumes, pollution concentrations) allowable limits.

Regulation can be focussed on one or more parts of the process. For example, if the problem is airport noise, a source regulation would require quieter jet engines. Pathway regulations would focus on airport operating hours and takeoff and landing paths. Receiver regulations would range from ear protectors for personnel, to building insulation, to land use zoning adjacent to the facility.

A critical factor in the success of any regulation is the level of our knowledge regarding the processes depicted in Figure 7.2a. The simplest instance of this approach involves the direct destruction of a receiver, entailing no pathways—an example being the construction of an apartment complex on a wetland. In GIS terms, this involves a spatial overlay in which a proposed activity is superimposed on the valued resource, as in Figure 7.2b. A regulation could prohibit the filling of wetlands, which are defined by an official map.

At the other extreme of complexity are the examples of acid rain in the Adirondacks or Greenetowne's sole source aquifer where multiple sources contribute loadings through complex pathway systems generating exposures to diverse receivers. Here our information and process knowledge base for regulations is severely limited. A fundamental deep information challenge to the regulatory process is knowledge building. The challenge is not just for large complex systems. The above wetland is not well protected by the ban on direct filling. Activities upstream in the watershed could lead to sediment loadings that over time will destroy it.

The mechanistic framing of environmental problems described above directly complements the administration of regulations by technocracies. Federal and state administrative procedures acts require standardized handling of permits and enforcement actions. Basic bureaucratic principles call for making decisions on standardized data by means of standardized "calculable rules" (Gerth & Mills, 1946). In the above simplified wetland example, the developer could be required to submit a site plan and a location map. The public technician could then plot the proposal on the official wetland map, and a yes/no decision would result. This is the most basic form of environmental modelling. The use of more sophisticated predictive models may provide a sounder basis for an informed decision. These approaches present a series of challenges which have been introduced in Chapter 4.

## REGULATION TYPES: SPECIFICATION AND PERFORMANCE

Historically, regulations took the form of prescriptive mandates. In medieval Venice, after some major fires, the munitions arsenal and the glassworks were moved to nearby islands. The regulation both identified the problem and articulated the form of the solution. In the nineteenth century, both the redeveloped Paris ("city of light") and Washington, D.C. set building height limits in large part to meet aesthetic objectives. By the late 1800s, Greenetowne and Browne County had enacted roadway design standards. They specified the width, maximum slope, and surface material and thickness for all new roads. The regulations applied both to the local governments, and to private developers who wanted to dedicate their new streets to the government as publicly maintained rights-of-way.

This regulatory approach constitutes a specification standard. The regulation specifies the location, design, and materials or equipment required for conformance. Initial predevelopemnt conformance is readily ascertained by means of simple map and document checkoffs from information already available from the developer. Subsequent in situ monitoring entails direct observations and measurements.

The specification approach was extended by Greenetowne to land use zoning in the 1930s. The purposes of zoning were twofold. First was the desire to protect the economic value of neighborhood real estate by keeping districts homogeneous. This was a response to serious tax base erosion and property abandonment which had occurred in New York and other large cities due to uncontrolled speculative construction (Boyer, 1983).

The second set of zoning objectives was related to the general state police powers of health, safety, and welfare. Based on national guidelines, zoning spatially segregated the location of various land uses to zones. In each zone there were unique requirements for minimum lot sizes, yards, and structural heights (Commerce, 1931). The basis for selecting these standards relied almost universally on the local judgement of appointed commissions who represented the power structure of the municipality. In major cities, the standards were influenced by studies prepared by professional designers. New York's pioneering concern with sunlight and daylight in response to rapidly increasing building height led to controls which generated Manhattan's famous midtowm "wedding cake" structures. For example,

> In a B district... an outer court or side yard at any given height shall be at least one inch
> in least dimension for each one foot of such height. (City of New York, 1916, Sec. 12).

The form of the regulation prescribed a set of horizontal distances and vertical heights, which were readily used by developers and checked by public inspectors.

An important new informational feature of this regulatory process was the "official map" which delineated the zones and accompanied the ordinance. The map played a critical role in the legitimization process wherein voters, elected officials, and judicial reviewers could visualize the comprehensive development plan for the community which the regulations were intended to achieve. Regulations called for a public hearing before the adoption of the map. Future revisions required the notification of adjacent owners, and a public hearing prior to a decision.

This official map was printed over a large-scale base map with typical district boundaries that followed readily identifiable bounds, such as street centerlines. The map was the sole means for informing property owners of the restrictions on their land and was readily available for inspection at the municipal building. It also served as a primary reference for monitoring and enforcement activities. As was the case with the roadway regulations, the information requirements for managing the specification zoning regulations were quite minimal.

In the decades between Greenetowne's adoption of the roadway and zoning requirements, the need arose nationwide to regulate building construction. Structural collapses, fires, boiler explosions, and the spread of communicable diseases such as TB and cholera had accompanied the increased urban densities. As in the case of zoning, the municipal response followed the lead of state and national initiatives. Some components of the building codes were simple specifications. For example, the establishment of maximum building heights was often based on the capacity of the local fire department's pumps and ladders. In contrast to the simple measurement-based standards of zoning and their intention of development homogeneity, the problems associated with buildings were performance-based, and the developers wanted maximum flexibility in design.

The innovative resolution was the development of "performance standards." These standards articulate one or more measurable criteria, and for each criterion, establish a performance objective, either a threshold (such as a minimum snow load structural bearing capacity) or a ceiling (such as maximum runoff discharge). The burden of proof is on the developer to demonstrate that the proposed project can meet the regulation in order to obtain the permit. Obviously, this approach can place enormous information burdens on both the private sector and the government. For buildings, much of this burden was accommodated over the two generations spanning the turn-of-the-century through the professionalization of the design professions and the generation of standardized manufacturing practices.

The major actors in this revolution are depicted in Figure 7.3. Universities developed standard courses of study in the design professions (primarily architecture and engineering) under the guidance of accrediting professional societies. These professionals learned standard practices and modelling techniques, and following graduation, supervised experience, and testing were licensed by new state boards to practice. Simultaneously, manufacturers joined in associations to standardize product sizes and specifications. These were cataloged and codified. Independent testing laboratories emerged to certify that products performed to a specified standard, such as load bearing or fire resistance. Associations of laboratories, manufacturers, and design professionals (such as the American Society of Testing Materials) emerged to coordinate testing standards.

By the 1930s, a modestly-trained building inspector could quickly certify that a development proposal conformed with the state building code by first making some simple drawing measurements and geometric calculations, and by certifying the construction drawings were signed by a licensed professional. The design professional in turn used a combination of professionally-sanctioned standard calculations (for example, structural analysis), and specified standard materials which are certified in perfor-

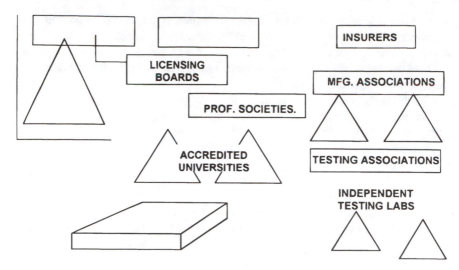

Figure 7.3. **Institutionalization of Building Performance Regulations**

mance by the independent laboratory used by the supplier. This approach serves both as an efficient delivery system and a framework for liability if and when a failure occurs. Public building departments kept copies of the approved plans on file for subsequent inspections and updating. The design professionals maintained copies of their calculations for the life of the project. This practice provided documentation of standard practice in case of a failure, which led to liability litigation.

This robust network is not static. Universities and manufacturers conduct research to develop new products and analytical techniques. With government sponsorship, this research periodically involves detailed in situ performance audits following catastrophic failures such as burst dams, hurricanes, and earthquakes. When a professional consensus on code change has emerged, the political process of revising the building regulations and educational approaches may proceed.

This system creatively merges the police powers of the public sector, the knowledge building orientation of academia, and the profit motives of the private sector. Regarding the latter, insurance companies have emerged to play a proactive role in the regulatory process. These firms insure the properties and lives involved, the manufacturers, and the licensed professionals. For example, in recent years the insurance industry has been a major advocate of upgrading building codes for hurricane resistance. Following major hurricane devastations in Florida, the industry pooled on-site data collected by claims adjusters and concluded that, despite the passage of strong state regulations, there were wide variations in their implementation by local governments. As a result the companies adjusted their rates accordingly, providing a strong message to jurisdictions and citizens regarding the performance of their public agencies (National Association, 1994). This approach represents a deep information approach to knowledge building.

Despite the early genesis and widespread success of the performance regulations in the building sector, some limits to its application may be ascertained. First, performance predictions must be made by the developer and confirmed by the agency. In the above discussion regarding buildings, the need and market was so large that the prediction process became independently self-sustaining. This resulted in reasonable available services to developers, and low costs to the government, particularly where professional certification served as a proxy for independent government predictions.

An example in which this system didn't initially work is interior daylight in buildings. With the large influxes of immigrants around the turn-of-the-century, and the high densities of crowded tenements, the incidence of TB and other communicable diseases rose dramatically. Over a period of years, New York, Boston, and other major cities passed a series of tenement reform specification standards that incrementally required increased yard space and the number of windows. Although health began to improve the specifications were not based on any predictive analysis.

During the progressive movement a major effort was undertaken to solve the problem by precisely controlling the amount of human sun and daylight exposure within residences. For a prototype application in a planned community, "Sunnyside," an elaborate set of graphic models and computations were developed to assess solar positions, the terrain, height, and shape of nearby buildings, window size and location, and interior floor plans (Heydecker, 1929). Although the specialists in charge succeeded in demonstrating the analytical basis for the regulatory process, it was entirely impractical for adoption. Neither colleges nor the design professions adopted the methods. Legitimization by a municipality would place a large unstandardized predictive burden on both the developers and the review staffs. Due to these implementation costs combined with the waning public health crisis, the effort to regulate a solar-human health performance standard failed, and building codes retained their simple rule-of-thumb specification standards of number and size of windows.

The adoption in the U.S. of solar and natural light performance standards would not take place until the energy crisis of the 1970s, when states like California passed regulations requiring access for solar collectors in new subdivisions. Rather than the complex challenge of human health, the prediction problem was highly simplified with the receiver being a solar collection building surface. An unobstructed direct beam path was required from a rooftop or wall to the sun during specified periods of the day, such as 9 a.m. to 3 p.m. (Thayer, 1979). Available prediction methods included graphical, scale models, and mathematical, using newly available computers.

In states adopting these regulations, the universities and professions rapidly accommodated their training and practice systems. An additional implementation factor involved the nature of the solar access regulation. Since access involved maintaining in perpetuity a wedge of open air space across adjacent parcels it was implemented as a set of easements in the deeds of the subdivision being created. Thus, landowners, lenders, and other interested parties had direct, continuous notice of the use constraint.

A second major challenge to performance regulation success involves the isolation of the systems that were regulated. As introduced in Chapter 4, all modelling involves simplifying assumptions. In systems analysis, a primary assumption involves the bounding of the system to be studied. For the performance-based building codes, each structure is independently analyzed and regulated. The developer just utilizes information about its own property and proposal. This simplification is valid because the major focus is on structural safety and interior functionality. In contrast, almost all recent environmental regulations focus on offsite impacts, and therefore entail much more extensive contextual data requirements for both the developer and the review agency.

A third limitation is our ability to predict performance. The central components of the performance building code are standardized computations and the use of standardized quality controlled materials. Of the potentially infinite arrangement of structural forms possible, only some can be comprehensively analyzed for structural performance. These are the forms that are taught and sanctioned. All materials are specified to perform as computed. These materials are manufactured by regional and national suppliers who adhere to the standards. Again, in contrast, once we leave the control of the building and go out on the site we encounter the "genius loci," the uniqueness of place. Geology, soil, vegetation, and weather are all highly variable and only partially predictable.

A final limitation is that building regulation, although based on national and state standards and actors, is locally implemented. The building reviews, inspectors, and record systems are within the local municipality. As we discuss below, with the emergence of state and Federally-implemented regulations, the communications distance between government and property owner may become a major barrier.

In the remaining portion of this chapter, we will examine some selected examples of environmental regulation. The format will be to briefly describe the origins and goals of the regulation, to visualize the information flow processes, and to critique the regulatory approach from the perspectives of environmental effectiveness and deep information.

## ILLINOIS STREAM CHANNELIZATION

Throughout the history of civilization, agriculture and development has taken place in flood plains. The availability of water, the level land, fertile soils, and ease of transport all reinforce this site selection process. Periodically, streams, rivers, and coastal areas flood. A traditional response has been the construction of dams to hold back water, and channelization projects to improve the rapid passage of peak flows. The latter includes the straightening, widening, and deepening of natural stream channels, as well as the removal of in-stream root "snags" and streamside vegetation. The hydraulically improved channels may be further enhanced by lining with concrete or stone rip-rap, and the construction of levees.

In the U.S., channelization (other than dredging of navigable waters) was historically not regulated. It has been estimated that between 1820 and 1970 over 200,000 miles of stream channels were modified, including over 34,000 that were channelized by Federal programs between 1940 and 1970. Over 80% of channelization has occurred in 15 states, which include Illinois (Mattingly, 1993). During the heightened environmental awareness of the 1960s and 1970s, channelization became a focus of environmentalists who opposed the widescale destruction of important habitat, particularly with the use of Federal funds. Public pressure and Congressional hearings eventually led to the elimination of the Soil Conservation Service's channelization program. Channelization projects proposed by landowners, such as farmers and developers, continue.

In Illinois, channelization is regulated by the Army Corps of Engineers (COE), and the Illinois Department of Transportation (IDOT). In addition, the Illinois Department of Conservation (IDOC) has broad responsibility for the environmental quality of riparian ecosystems. Many small projects, primarily agricultural modifications of minor tributaries, are exempted by national and state "generic" impact statements. Therefore, no public records are kept of these activities.

The major information flows of the permit process are shown in Figure 7.4. For regulated projects, the developers submit plans and cross-sections of the modification proposals. For efficiency this is a joint application to the COE and IDOT. Engineers in these agencies use numerical hydraulic models to check the capacity of the proposed channel to carry flood flows. Environmental concerns can be introduced in subsequent reviews by agencies such as the U.S. Fish and Wildlife Service and IDOC. The former rely heavily on their 1" = 2000' national wetlands map series discussed in Chapter 9. IDOC utilizes a variety of internal resources, including a database, the Illinois Stream Information System (ISIS), which is based on the USGS quad maps, and more recently a stream habitat classification system. Field investigations and public hearings may be part of the process.

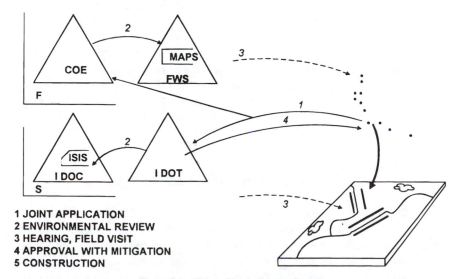

1 JOINT APPLICATION
2 ENVIRONMENTAL REVIEW
3 HEARING, FIELD VISIT
4 APPROVAL WITH MITIGATION
5 CONSTRUCTION

**Figure 7.4. Illinois Channelization Permit**

Most applications are approved. Often, based on the opinions of the review staffs and in consultation with the applicant, conditions are attached to the permit covering erosion control and revegetation. Unlike the calculated predictions of channel capacity, which constitute a performance standard, these mitigation conditions are professional judgement-based specification standards.

This permit system is a classic example of disjointed incrementalism. Over decades, hundreds of landowner-initiated projects are individually reviewed and authorized. In a major environmental performance audit, a team from the University of Illinois attempted to assess the environmental outcome of the permit program. Their conclusions were deeply troubling. Some of the primary results were:

1.  Under the permit system the riparian habitat is being systematically destroyed;
2.  Much of the cumulative damage is uncontrolled due to the small projects which are waived from the permit process;
3.  Neither the information provided by the applicant nor available internally by the agencies adequately describe the existing environmental characteristics, so no record of the "before" ecological condition is available for subsequent analysis; and
4.  Regarding permit conditions, their case study field checks revealed that "even this minimal approach to mitigation is not implemented." (Mattingly, 1993, p. 789)

Clearly, the disjointed agency information system represents an efficient traditional engineering performance review core subsystem, with an ad hoc collage of shallow environmental review subsystems tacked on.

From all three "deep information" perspectives, the channelization permit process is a failure. There is little or no shared knowledge building of the hierarchial environmental processes. The richly complex ecosystems of the riparian corridors are ignored in the data collection permit process, and in the narrowly-focused stream fisheries database. The information process is not open. The investigators had to go to great lengths to try to assemble a comprehensive picture of the resource destruction from a number of isolated data sources. Local governments and affected communities are not informed of the cumulative impact of the permit program. Responsible landowners are not educated or monitored regarding their continuing obligations, and communities are not informed of landowner violations.

## NEW YORK STATE WETLANDS

In the Illinois channelization example, the jurisdiction of the regulation is "floating." It is grounded wherever a property owner proposes a stream modification that isn't exempted by a generic approval. The other primary mode of establishing jurisdiction is by means of an official map, like Greenetowne's zoning ordinance. In this approach, the legislature mandates not only the control of an activity but also the responsible agency to spatially delineate where the regulated phenomena exists. The official map is formally adopted prior to the start of implementation.

In the 19th Century, wetlands were seen as having a set of negative values ranging from harboring disease to constraining agricultural productivity. Massive governmental and private efforts were undertaken to fill and drain the "swamps." By the late 20th Century, the Federal government and many states had reversed their values and passed regulations protecting the positive functions of wetlands, which included habitat, flood protection, aesthetics, and water quality.

The legitimization of wetland protection involves two primary steps, the classification standard of what is a protected wetland, and the map delineation of the regulated areas. The debate regarding classification is ongoing, particularly at the Federal level. Since a wetland includes a dynamic mixture of (seasonal) surface water, soil, and vegetation, and since the U.S. has such geomorphic and biological diversity, the basis of this high development stakes decision is easy to understand. This classification issue is further discussed in Chapter 9.

Two different approaches to official mapping have evolved. At the Federal level, the U.S. Fish and Wildlife Service has relied on remote sensing (air photo and satellite) to map wetlands based primarily on surface water and gross vegetation characteristics. In contrast, New York's wetland regulations, promulgated by the Department of Environmental Conservation (DEC), classify on the basis of detailed plant species, thus necessitating in situ delineation. The vegetation types and associated water regimes, such as "kettlehole bog," provide the basis for an importance hierarchy of four classes that is used in evaluating permits.

New York was a pioneer in the protection of wetlands, passing its Freshwater Wetland Act in 1975 (N.Y. Environmental Conservation Law Article 24). It regulates wetlands that have a surface area equal to or greater than 12.4 acres (approximately five hectares). This threshold was established by professional judgement that considered the nature of the resource and the practicalities of mapping the entire state. In the late 1960s, the State's Office of Planning Coordination (now defunct), responding to growing concerns regarding the effects of massive suburbanization, developed one of the first statewide systematic natural resources inventories. The Land Use and Natural Resource inventory (LUNR) study used air photo interpretation to produce classification overlays for the 1' = 2000' USGS quadrangle base maps. The inventory had 100 categories, including forested and open wetlands. The air photo analysis provided a comprehensive overview of the resource, and when compared to the wetland features shown on the older USGS quads, clearly demonstrated the rapid losses in many regions.

Adoption of the official maps has three basic stages: map preparation; notification; and approval. These are shown in Figure 7.5. Using the LUNR overlays as a framework, the intensive field identification of vegetation types was undertaken on a county-by-county basis in a broad cooperative process involving government staff, colleges, and environmental NGOs. Field notes regarding plant species, terrain, and water features were recorded for each wetland site. Preliminary maps including a unique identification number, wetland class, and wetland outline were delineated on all the quad

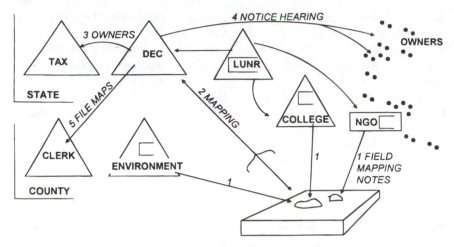

**Figure 7.5. N.Y.S. Wetland Map Approval Process**

sheets for a county. Once the preliminary maps were developed, the informational process which developed decades earlier for zoning was followed.

> Upon completion of the tentative freshwater wetland map for a particular area, the commissioner... shall hold a public hearing... the commissioner shall give notice of such hearing to each owner of record as shown on the latest completed tax assessment rolls. (N.Y. Environmental Conservation Law Article 24, 1975)

As each county's preliminary mapping was completed, state staff cross-checked the maps with the official tax maps, and sent a notice to each potentially affected property owner. This was followed by a public hearing and an opportunity to modify the maps prior to adoption. Once approved by the commissioner, a set of official maps was kept in regional state DEC offices, and another set was deposited with the County Clerk. Shortsightedly, following the one-time mandated notification, most of the owner's parcel numbers were thrown out by the agency. Similarly, the original field notes that described the in situ details of the resource were often not kept.

Implementation of the permit system is shown in Figure 7.6.

First, a prospective developer needs to find out whether its property may contain a wetland and if its project may require a permit. This is highly ad hoc. If the project requires other state permits, the state staff systematically checks the official maps. If the project requires local permits, a local building inspector or town planning board may sometimes signal the wetland issue. If the project doesn't require another permit, the owner may proceed to bulldoze the affected area in complete ignorance.

As was shown in Figure 4.4, the official wetland maps at 1' = 2000' are inappropriate for a detailed review of a proposed site design. This was the same problem Illinois had. In addition, the regulated zone includes the mapped wetland and a 100'

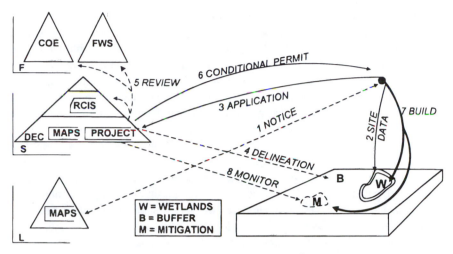

**Figure 7.6. N.Y.S. Wetland Map Permit Process**

buffer around the wetland. The buffer constitutes a highly simplified recognition that activities in proximity to a wetland, such as construction-generated erosion, can impact the resource. The applicant must provide a quad sheet location map and project plans, including building and earthwork, at 1' = 50' or larger, and clear photographs of the site and wetlands area. Since the official map is too gross to base most decisions on, it is necessary to re-delineate the wetlands in the field. This may be done by DEC staff or by a private consultant (NY Dept. of Environmental Conservation, 1991).

Once filed, the basic description of the application is entered into DEC's Regulatory Compliance Information System (RCIS). RCIS is a single-purpose, depart-ment-wide Uniform Procedures permit tracking system designed to facilitate the time-ly processing of new and renewed permits (NY Dept. of Environmental Conservation, 1987). It is not designed to act as an environmental information system. There is no use of Parcel Identification Numbers, nor in the case of wetlands does it contain the wet-land identifier, its class, or any descriptive information regarding the size, characteris-tics, or functions of the resource.

Decisions regarding filling of wetlands involve the weighing of the environmen-tal values with the "beneficial economic, social and agricultural development of the state" (article 24, section 103). In addition, the DEC avoids "taking" controversies wherein a landowner argues in court that the denial of a permit removes all feasible economic use of a parcel and therefore constitutes a constitutional taking that requires state compensation. Most applications are eventually approved, typically with mitiga-tion conditions. These may range from construction practices to the development of a new wetland to satisfy the Department's "no net loss" policy.

The analysis and approval process rarely utilizes performance predictions. Although the law specifies nine public benefits from wetlands, ranging from flood control and education to food cycle nutrients, most decisions are based on professional judgement and socio-economic pressures. The department frequently checks with other state and Federal agencies (there is a joint application process with the COE). Individual hard-copy project files are created for each permit. Once permitted, development can proceed. Monitoring of the development and the required mitigation is optional. The permit is a one-time approval with no renewal requirements.

From the three facets of "deep information," the wetland permit program is a mixed success. The initial mapping process, which involved a rich variety of community efforts, was a solid start to shared knowledge building of the resource. However, once implemented, the permit process essentially ignores the hierarchial complexity of wetland systems and processes. No progress has been made on transforming the simplistic 100' buffer specification standard to a meaningful performance requirement. Requirements for creating new ameliorative wetlands are not well monitored or managed. The open system which was possible at the outset with landowner notification and mapping has fallen into disarray due to inability of the agency to develop a Land Information System which would be transactionally updated.

Downstream end users remain ignorant of their continuing responsibilities. In stark contrast to the public sector one-stop agency wetland permitting is the Natural Area Registry Program of the state's branch of the Nature Conservancy. To complement its traditional ownership acquisition of bio-diverse areas emphasis, the Conservancy developed a voluntary registry program. Landowners of unique resources, many of which include wetlands, are offered free analysis of their holdings and stewardship recommendations, including information to help them assess the health of their ecosystems. Services include periodic site visits and a newsletter (Nature Conservancy of New York, 1993).

## WATER AND AIR DISCHARGES

A serious information shortfall in both the channelization and wetland examples is the lack of effective post-construction monitoring. In effect, the permit represents an open-ended one-stop license. Although this practice of benign neglect has long been common for permits whose primary impacts are related to construction activities, a different approach is standard for activities which cause continuing impacts due to process discharges. For these air, water, and land disposal activities there is typically three stages of review: construction approval based on performance modelling; start-up permitting based on in situ performance measurements; and periodic (2–5 year cycles) repermitting. Two critical information dimensions of this multi-stage approach are the simplification assumptions of the environmental model, and the design of the monitoring programs.

Greenetowne's initial involvement with environmental management was focused on cholera, a water-borne disease. It is difficult to directly detect pathogenic bacteria in a water source due to their minute presence in huge volumes of water. This problem increases by orders of magnitude for viruses. A pragmatic alternative involves the use of a surrogate. For water quality, *E. Coli*, an intestinal bacterium which all of us daily excrete in tremendous numbers, has long been used in monitoring. It persists in the environment, is easy to culture in a lab, and its presence strongly indicates the potential for human waste in the water. The Browne County health department samples the Blue River and the public well supply daily for *E.coli* and a number of water quality chemicals.

A second Greenetowne water crisis involved the eutrophication of the Blue River. This process has some human health and aesthetic impacts but a primary outcome is often massive fish kills. As Federal and state governments got involved in regulating pollution discharges from public and private sources, the aquatic "health" of the surface waters became a public goal. Translating this goal into measurable criteria and functional objectives represents a considerable challenge. Early in the century, the emerging water pollution technocracy legitimized dissolved oxygen (DO) as an efficient indicator of stream health. Natural waters are commonly saturated in DO due to the generation of oxygen by aquatic plants and the surface mixing of water by wind and waves. As dissolved oxygen levels increase, a broader range of fish is supported at the top of the food chain. In particular, trout, a major target of sport fishing and its public and private industry, require high levels of DO.

Many organic pollutants, such as untreated sewage and some chemical pollutants accelerate oxygen consumption by waterborne organisms (biochemical oxygen demand, BOD). When this rate of consumption exceeds the natural rate of reaeration, the level of DO "sags." If it sags to zero, the aerobic organisms die off and are replaced by anaerobic organisms. Therefore, water quality regulations were initially designed to classify the "best" use for a water body in an official map and to specify performance standards, such as the threshold allowable level of DO which must be maintained.

A classic case study of the utilization of early modelling methods to make water quality regulatory performance decisions involved the Delaware River Basin Commission's efforts to restore the highly polluted reach from Trenton, NJ, to below Wilmington, DE, a zone with over 40 major discharges. DO was identified at the start of the study as the major indicator to be used. Measurements along the river established the existing DO sag profiles and estimated the oxygen demand pollution loading of the primary sources.

Input-output, deterministic computer models were developed for thirty discrete subreach chunks of the study zone. Model calibration was done to fit the predicted mean DO values to observed values obtained from monitoring. This fit is shown in Figure 7.7. Alternative waste treatment plans were modelled that would raise the average expected DO from 1.0 ppm (parts per million) to 4.5 ppm. Cost-benefit estimates of these plans and the projected DO outcomes provided the basic decision information for the elected officials and regulators.

The authors of the case study of this process have two fundamental criticisms of this environmental performance regulatory approach. First, the deterministic modelling of the river and its reliance on average values highly oversimplified the system dynamics of this complex estuary.

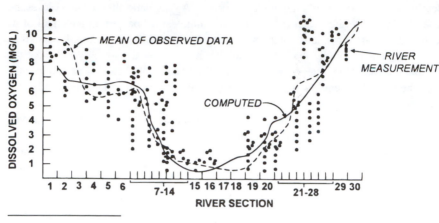

ᵃ based on Ackerman, p. 58

**Figure 7.7.  Predictive Computer Model**ᵃ

The principal conceptual obstacle that rivers like the Delaware (below Trenton) pose to sanitary engineers is that they are estuaries and are therefore influenced by ocean tide. This means that BOD flows not only downstream but upstream as well, complicating all calculations immensely. (Ackeman et al., 1974, p. 33)

In addition to the spatial chunking of the river into thirty zones, and the complexity of tidal flows, the modelers needed to simplify time due to the tremendous variations in rainfall, runoff, and temperatures which could occur during the lifetime of the proposed treatment facilities.

To reduce the data base and computational resources required for a 'time varying analysis' the [staff] primarily attempted to predict the DO curve on the hypothesis that the *relevant river conditions remained constant over time* ... the 'steady state' approach. (p. 37)

From a visual inspection of Figure 7.7, it is clear that the mean value is a poor approximation of the actual condition at any given time.

The second main critique focused on the extreme reductionism by which a highly complex natural and socio-economic system was reduced to a single easily measured criteria and performance objective, DO.

a decision maker would be extremely myopic if he were to adopt DO as the exclusive touchstone of water quality. [It] has little impact on human water consumption so long as anaerobic conditions producing serious odor problems are avoided.... [for] DO is a far from sufficient indicator for swimming, fishing or boating.... Nor does the DO profile provide an adequate basis for determining the impact the Delaware has in the larger ecological balance. (p. 27)

Prior to 1972, the Federal government had relied on states to classify waters and to regulate via performance-based permits. The system proved unfeasible due in large part to the complexity of enforcement of violations due to a particular discharge in a receiving body system which had many sources. The 1972 Federal Clean Water Act (33 *USC* 1251) shifted from a systems to a site performance level. These "command and control" regulations set the allowable levels for each discharge. The law specified that by 1977 all sources had to be upgraded to "best practical control technology available", and by 1983 "best available technology economically achievable". Amendments created the National Pollution Discharge Elimination System (NPDES) of permits.

Although our ability to model complex systems has improved considerably since this study, the waste water discharge permitting process remains highly reductionist. Referring to Figure 7.2, there are three interrelated systems which need to be understood: the sources, pathways, and receivers. In the Delaware study, all of the modelling effort was focused on the river as pathway. No systems views were analyzed for the sources, such as the internal processes of a municipal sewage treatment plant, or of the ultimate receivers, such as a risk analysis of the ecosystem.

The majority of environmental information which enters permit-based regulatory systems is second party self-monitoring. An applicant for an NPDES permit must provide a variety of detailed descriptive and predictive information to the state or Federal review agency. For example, this requires a schematic model of the facility's water mass-balance flows. Laboratory tests must be conducted by the applicant to identify the amounts of a "toxic pollutant" (e.g. asbestos), and approximately three hundred listed "hazardous substances." Effluent must be reported as concentrations and as total mass. For continuous discharges, flows are to be reported as daily averages. For "intermittent" or "seasonal" discharges, flows are reported as "days per week," "months per year," "long term average," and "maximum daily" (EPA, 1985).

Although this is a Federal regulatory program, most states have been authorized by EPA to review and issue permits. The detailed application, its accompanying engineering reports, and state staff analyses are kept at the state level in a hard-copy project file. Summary data on each permit issued are forwarded to EPA for inclusion in its facility database system, FINDS. When disseminating this data, EPA includes its filter for confidentiality.

> Information contained in these application forms will, upon request, be made available to the public for inspection and copying. However, you may request confidential treatment of certain information which you submit on supplementary forms…. No information on Forms 1 and 2A through 2D may be claimed as confidential. (EPA, 1980, p. 1–2)

The Clean Air Act (42 *USC* 7401 *et seq*) permitting process works much the same way. Initial permitting is based on modelling analyses of the regulated emissions. Following construction, a testburn trial period involves in situ monitoring by the regulating agency to establish whether the facility is working as designed. If satisfactory, a renewable permit is issued.

One of the many simplifying assumptions made in the Delaware example was that, once regulations were established and permits issued discharges would conform to the specified limits. In their groundbreaking study on enforcement, Russel et al. reject this assumption as naive. They cite studies which find,

> a significant fraction of air pollution sources are out of compliance for substantial parts of every year and that failure to comply with permit (wastewater) discharge limits is wide-spread…. Achievement of initial compliance usually has no effect in itself on environmental quality. (Russell, 1986, p. 6-8)

Clearly, monitoring throughout the operating life of the facility with regulatory enforcement of standards violations is a prerequisite for achieving the environmental performance objectives of the permitting program.

The development of a monitoring information program involves answering the basic interrogatives who, where, when, and how. Most monitoring involves self-reporting by the permittee. Although this is very efficient for the government, it has some serious limits. Not all operators maintain quality sampling programs and not all report findings honestly. Therefore, the regulating agency must play a periodic oversight role. Spatially, there are two distinct systems to be monitored: the source and the receiver. For the facility as source system, monitoring typically entails collecting samples at the stack and pipe outlets. If the efficiency of pollution treatment processes is in question, then samples must be taken at intermediate points within the facility.

For regulators, a basic information policy issue involves the right to enter a private facility in order to monitor. Although case law regarding constitutional search and seizure protections is mixed,

> there has been a consensus among the circuits (courts) that a warrant is required under the Clean Air Act. (p. 70)

The giving of advance notice for an in situ monitoring inspection significantly limits the ability to find violations. Technologically, advances in permanent continuous monitoring equipment built into the facilities (some with direct electronic reporting to the regulator) and remote offsite monitoring systems may eventually eliminate this problem.

Controlling discharges is a means to achieving a desired set of ambient levels in the receiving system, such as a lake or airshed. Ambient monitoring downstream or downwind of facilities provides critical information about both the effectiveness of the regulatory system and the quality of the modeling on which the construction permits was founded. Long-term ambient monitoring is chronically underfunded. For major emitters, the conditions of the operating permit may include the support of some offsite ambient monitoring. Most of this effort is the responsibility of government. For example, in the Syracuse metropolitan area there is one air monitoring station, the minimum needed to report data to the state and EPA. In Chapter 5, the use of trained NGO monitors, such as Lake Associations, was described as one approach to data generation. In general, however, we often have insufficient information to understand the functioning of the receiving system.

The timing of monitoring samples represents a major challenge. Most facilities have highly dynamic operating conditions which quickly respond to differences in inputs (a sewage treatment plant) or differences in demand (a manufacturer). Therefore, we would expect the volume and concentration of emissions to be variable. As described above, discharge permit applications require various estimates of discharges. In order to be statistically valid, a monitoring program must incorporate a temporal sampling design which addresses these fluctuations. Even in a well-implemented program there may be false "findings" of violations, and false certifications of actual violators. Because of this underlying uncertainty, agency decisions to proceed with enforcement and subsequent judicial conclusions have varied.

The implementing agency has major discretion regarding where and when to monitor, and whether or not to enforce if a violation is found. The realities of operating complexities, agency resource limitations, and political will mean that some fraction of facilities will always be nonconforming. One tactical approach to this problem comes from management science, whereby the agency attempts to maximize the return on its investment by analyzing the probabilities of finding and correcting violations and the magnitudes of the environmental payoffs in reduced pollution.

A different view of monitoring has emerged from the environmental justice movement. Over the last decade a number of surveys have shown that agency enforcement tends to be weakest in communities that lack political power. One key to effective monitoring and enforcement traditionally has been for those affected to pressure their elected officials to influence the agencies to act. The poor and minority groups are typically not effective actors in this process.

From the perspectives of "deep information", air and water discharge permits leave much to be desired. Regulations are often based on highly simplistic system assumptions. Once in place, the regulatory programs quickly shift focus to implementation and these initial approaches are not modified by a systems learning feedback process. Traditionally, the permit information processes have not been open and plural. Some participation has occurred through citizen monitoring programs. The fragmentation of regulations and agency responsibilities has presented major obstacles to downstream information users. Even with the FOIA, it was virtually impossible for a community to assemble a composite picture of its pollution load. The Community Right to Know Act discussed in the next chapter has made considerable progress in addressing this problem. Because state and Federal regulatory agencies rely on generalized "facility" spatial descriptions rather than tax-based property ownership PINs, it is extremely difficult to integrate permit information into a LIS.

## SUMMARY

Environmental regulations emerged in the post-WWII period as the primary means for reducing pollution and protecting natural resources. The history of the regulatory movement has been one of increasing regulation accompanied by increasing administrative fragmentation. The regulations are the largest generator of environmental data. The majority of this data is highly localized, second party self-reporting. Its primary format is the hard-copy application file, which includes permit forms, maps, technical reports, and agency correspondence. Typically, only selected summary data ever gets into a digital format.

Information technology can help overcome these limitations. A recent pilot study in the Adirondack Park Agency focused on full text scanning and indexing of project files, and linking them to digital taxation land ownership records through the use of parcel identification numbers, which the Agency requires in all permit applications. Building on a process developed to develop a modern LIS for the extensive records of the Federal Bureau of Land Management, the pilot study demonstrated the feasibility of transcending narrow MIS permit tracking approaches to including the entire text and map contents of the traditional project files. An additional component of the pilot was to make this information accessible to distributed municipal sites (Center for Technology in Government, 1995). This type of major reengineering is essential if we are to disseminate substantive environmental information and improve regulatory agency accountability.

Regulations are often based on weak understanding of the affected systems, yet the vast majority of the data generated is not focussed on the development of improved systems information and knowledge. The pragmatic emphasis in pollution control on major point sources conveniently downplayed the importance of receiving systems. Cumulative impacts are typically ignored in a process characterized by disjointed incrementalism. The primary focus of regulatory information systems development has been to increase the efficiency of the agencies and reduce the sometimes onerous self-reporting burdens on the permitees.

In recent years the larger systems scale concerns have reemerged. Examples include the old growth forests of the northwest, Remedial Action Programs in indus-trialized embayments of the Great Lakes, and the sole source aquifers, wellheads, and surface water supplies now regulated by the Safe Drinking Water Act (42 USC 300).

The quality of regulatory data as a basis for environmental understanding is often suspect due to its second party sources, the variation in monitoring, and the ad hoc location and timing of developer-initiated records. No one would believe a "state of the wetlands" report issued by a state commissioner that is solely based on a tabulation of the department's regulatory permit files listing acres filled, ameliorated, and created. Acreage is a poor proxy for systems functions, many activities go unregulated, and many permits are not implemented as approved.

In Chapter 8 we will focus on the role of information as complement and supple-ment to regulatory policy. In Chapter 9, the policy basis of scientific information sys-tems is explored as a means of developing a comprehensive action-oriented understanding of our environmental resources.

# REFERENCES

Ackeman, B. et al. *The Uncertain Search for Environmental Quality*. New York: The Free Press, 1976.

Boyer, M. *Dreaming the Rational City*. Cambridge, MIT Press, 1983.

Center for Technology in Government. *Balancing Environmental Quality and Economic Viability in the Adirondack Park*. Albany, NY: SUNY, 1995.

City of New York. *Building Zone Resolution*. New York: City of New York, 1916.

Gerth, H. & C. W. Mills. *From Max Weber: Essays in Sociology*. New York: Oxford University Press,1946.

Heydecker, W. "Sunlight and Daylight for Urban Areas," in *Regional Survey of New York and Environs*. New York: Regional Plan Association, 1929.

Mattingly, R. et al. "Channelization and Levee Construction in Illinois: Review and Implications for Management, *Environmental Management,* 17 (6) (1993): 781–795.

National Association of Independent Insurers. Mitigating Catastrophic Property Insurance Losses. Des Plaines, Ill, The Association, 1994.

Nature Conservancy of New York. *The N.Y. Natural Areas Registry.*

NY Dept. of Environmental Conservation. *Freshwater Wetlands Program: Applicant's Guide*. Albany, NY: The Department, 1991.

NY Dept. of Environmental Conservation. (1987). *User Guide—Regulatory Compliance System*. Albany, NY: The Department, 1987.Rochester, NY: Nature Conservancy of New York, 1993.

Russell, C. et al. *Enforcing Pollution Control Laws*. Washington, D.C.: Resources for the Future, 1986.

Thayer, R. *Solar Access: It's the Law—A Manual on California's Solar Access Laws for Planners, Designers, Developers, and Community Officials*. Davis, CA: University of California Appropriate Technology Program, 1979.

U.S. Department of Commerce. *The Preparation of Zoning Ordinances: A Guide for Municipal Officials and Others in the Arrangement of Provisions in Zoning Regulations*. Washington, D.C.: The Department, 1931.

U.S. Environmental Protection Agency. *Application Form 1-General Information: Consolidated Permits program*. Washington, D.C.: USEPA, 1980.

U.S. Environmental Protection Agency. *Application Form 2C- Wastewater Discharge Information*. Washington, D.C.: USEPA, 1985.

# 8

---

# Information as Self-Regulation

......................

## INTRODUCTION

*I*n the previous chapter we examined the role of police power based, command and control regulations as a means of protecting environmental processes through direct governmental limitation of public and private sector activities. This approach has achieved major reductions in pipe and stack emissions, but as we have seen it has several major intrinsic limitations. Environmental protection is also an individual activity of the citizen or landowner, and a social activity involving various levels of community. In her eloquent analysis of urban revitalization, Jacobs has said:

> The first thing to understand is that the public peace—the sidewalk and street peace—of the cities is not kept primarily by the police; necessary as the police are. It is kept primarily by an intricate, almost unconscious, network of voluntary controls and standards among the people themselves. In some city areas... keeping of sidewalk law and order is left almost entirely to the police and special guards... such places are jungles. (Jacobs, 1961), p. 31.

Just as information plays critical roles in the development and implementation of regulations, so it is essential to the development of individual behavior and the shaping of group action. A re-examination of Figure 5.2 shows that the regulation information flow is typically a two-actor dialog between the agency and the landowner. More robust schemas involve broader sets of actors and stakeholders.

A measure of effective communications is not just that a message has been received and properly interpreted, but that it has modified behavior. In this chapter, we will briefly examine three arenas in which information policies are designed to complement, and in some situations serve as an alternative to, regulations. The first arena is the real estate marketplace and its associated lenders and insurers. The second arena is the community, which contains regulated facilities. These operations directly affect its infrastructure and supporting services, and its citizens who are potential receivers of emissions from normal activities and emergency releases. A third information policy arena involves government provision support programs where "best management practices" (BMP) by the landowner as a condition of receiving support are seen as an alternative to direct regulation.

## HAZARDOUS WASTE AND THE REAL ESTATE MARKET

In Chapter 6, we saw that the emergence of the cadastral public information system was a direct response to a market failure, the inability to efficiently ascertain who owned what land. The system that was developed was highly effective, in large part because it focused on spatial location and owner lineage, and used broad concepts like fee simple to avoid describing the complex environmental attributes of the land. After centuries of operation, the cadastral system has become deeply challenged by the aftermath of Love Canal.

Congress' response to Love Canal was unique in the evolution of environmental regulations. In the 1980 Comprehensive Environmental Response, Compensation, and Liability Act, "Superfund" (CERCLA, 42 USC 9601) and its Superfund Amendments and Reauthorization Act of 1986 (SARA), Congress not only imposed controls over future activities, but assigned joint and several liability to past activities including those that were legal at the time. In essence, any former landowner could be held liable for the cleanup of newly designated hazardous wastes. The full impact of this revolutionary regulatory approach came into focus for the real estate community in *U.S. v. Maryland National Bank and Trust Co.*, where a bank which had gained title through a mortgage foreclosure was held liable for a costly cleanup (632 F.Supp. 573 (D. Md. 1986)).

Suddenly all real estate transactions, particularly those involving former commercial and industrial properties, were under a potential cloud of hazardous waste remediation liability for any former activity on the site. In the SARA amendments to CERCLA, Congress provided a third party defense to liability which has become known as the "innocent landowner" provision.

> SARA added a new definition to Superfund—that of 'contractual relationship'—which excludes from liability those transfers of ownership or possession which occurred after the disposal occurred, provided the transferee can prove... that it did not know, and had no reason to know, of the existence of a hazardous substance on the site... SARA requires that a purchaser undertake appropriate inquiry into the condition of the property... Congress was quite specific on this score, specifying that the specialized knowledge and experience of the purchaser, the relationship of the price to the value of the contaminated property, any commonly known or reasonably ascertainable information about the site, and how easily the contamination could have been detected must be taken into account. (Turner, 1988, p. 6)

Clearly the traditional cadastre contained little of the descriptive information required by the innocent landowner defense. This vacuum quickly spawned an entire industry, voluntary private sector environmental site audits. Typically an initial (Phase I) audit consists of four components:

A.  A review of public hazardous waste site related records, such as the EPA's National Priorities List, and state remediation lists for a one mile radius around the site;
B.  A review of other publicly available records, such as zoning and fire insurance

maps for the property and its adjacent parcels;

C. An onsite visual investigation of the property and adjacent parcels looking for fuel storage, drums, lagoons, etc.; and

D. If the site is currently in use, interviews with workers regarding present and former operations. (American Society for Testing Materials, 1993)

This process is shown in Figure 8.1. As a private contract, the product of the investigation is a report to the client.

Almost every aspect of this critical information process has been the subject of debate. Basic issues include standardization, professionalization, public control, and public access. Although the Federal government has audit procedures for Federal properties, in general private audits are ad hoc and unregulated. For example, in New York City, most major banks have developed their own internal guidelines for audits which precede real estate financing. In recent years, both the American Society of Testing Materials and the American Society of Civil Engineers have promulgated audit standards. The former is being widely adopted in the public and private sector, while the latter focuses on Phase II audits which entail the taking of site samples with subsequent in situ and laboratory physical testing. There is currently no licensing of site auditors. Auditing services are offered by lawyers, engineers, and environmental consultants. In addition, a number of information industry firms have emerged which provide digital GIS-based access to the many of the records needed in a Phase I audit.

The identification of potential hazardous waste locations is a public as well as a private responsibility. State and Federal facility permitting record series, such as EPA's FINDS database, are a primary source of identification. As discussed previ-

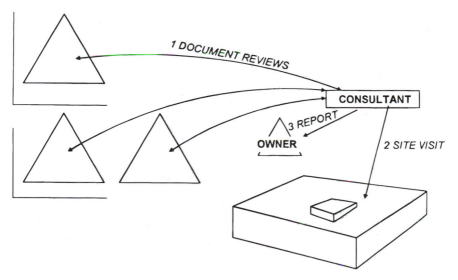

**Figure 8.1. Private Sector Site Audit**

ously, these records often have generalized location descriptors, and rarely are connected to parcel ownership records. The EPA and states have developed lists of known and potential sites, and "Potentially Responsible Parties" who might be liable for cleanup. Some states, including Minnesota, West Virginia, Pennsylvania, and New York, have acted to formally connect these lists to the cadastre. For example, the New York regulation requires a state-generated list of sites and their PINs be available at each county clerk's office.

Since most site audits are private transactions it can only be presumed that many potential problems identified in Phase I investigations are never made public and never cleaned up. New Jersey has taken a much more proactive approach to auditing and remediation. In its Environmental Cleanup Responsibility Act (N.J.S.A. 13:1K6 et seq), prior to the sale or shutdown of an industrial facility a site audit must be conducted, and

> either a negative declaration be prepared or an adequate cleanup plan be prepared and implemented... In essence, it demands proof that if hazardous materials were handled and problems arose, they either have been or will be cleaned up. (Highland, 1987, p. 23)

In this case, the audit information and the associated remediation plan, although generated by private contract, is mandated second party self-reporting to the state regulator (unless the owner decides not to sell or close). The information flow is shown in Figure 8.2.

In addition to playing a critical role in sales of real property, audits can provide a comprehensive framework for the day to day and strategic operation of a facility. A Price Waterhouse survey concluded that three-quarters of companies, including almost all large companies, have conducted corporate environmental audits. Reasons given for

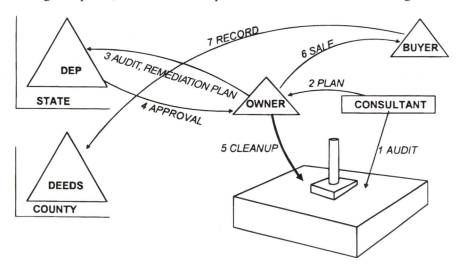

**Figure 8.2.  NJ industrial Facility Site**

this voluntary practice range from more effective compliance, and increasing competitiveness, to reducing Superfund liability, and meeting disclosure requirements of the Securities Exchange Commission (Kass & McCarroll, 1995).

A fundamental policy issue is whether the results of such audits should be made public. A corollary issue is whether a company should be penalized for voluntarily disclosing pollution violations. EPA has issued a policy on "Voluntary Environmental Self-Policing and Self-Disclosure" which attempts to strike a balance by requiring audits to be made public information while limiting punitive enforcements. A further policy complication comes from the 18 states which have passed "privilege statutes" to protect internal voluntary audits. In response, the EPA threatened to rescind its delegation to the states of Federal pollution enforcement statutes.

## COMMUNITY RIGHT TO KNOW

The 1970s was a decade of increased concern over public health, including health in the workplace. Starting with voluntary Federal standards, which were followed by mandatory reporting requirements in some states, workers and labor union pressure finally resulted in The Occupational Safety and Health Administration (OSHA) adopting a Federal Hazard Communication Standard in 1984. This standard required employers to collect information regarding chemicals and inform workers via standardized Material Safety Data Sheets (MSDS).

The 1984 Bhopal, India chemical disaster, which was followed in 1985 by a major leak in West Virginia, raised public awareness in the U.S. that communities were poorly prepared to deal with industrial accidents. Many fire departments and other local emergency response groups were ignorant of the specifics regarding hazardous processes, materials, and locations within industrial sites. The 1986 SARA included a stand-alone Title III, Emergency Planning and Community Right-To-Know (EPCRA).

EPCRA is a major example of informational federalism. It has two primary thrusts. First, it establishes a process intended to inform and prepare local emergency groups to effectively handle industrial disasters. Second, it creates a national data set of annual chemical releases (permitted and accidental) to the environment, the Toxic Release Inventory (TRI), discussed below.

Under EPCRA, each state is to create State Emergency Response Commissions (SERC). The Commissions establish Local Emergency Planning Committees (LEPC). Industries which handle designated materials above a threshold amount, are required to notify the SERC and LEPC (note: the somewhat inconsistent classification by EPA of materials on different regulatory lists, and the continuing revision of these lists, is a matter of ongoing debate). Regulated industry and commercial establishments are subsequently required to provide detailed information regarding material types, quantity, and specific on-site locations. In addition, they must provide appropriate MSDS sets to the SERC, LEPC, and the fire department. The public may request this information, with the LEPC making the filter judgement between the need for public safety and the

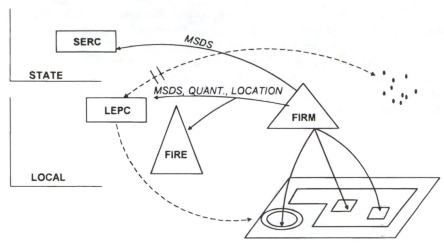

**Figure 8.3. EPCRA Reporting**

owner's confidentiality. The LEPC is authorized to make announced site visits. When a spill occurs, there are additional specific notification requirements. These information flows are shown in Figure 8.3.

The mandatory Federal requirement of local informational committees is quite unique. The law is very specific about the composition of the LEPCs:

> Each committee shall include, at a minimum, representatives from each of the following groups or organizations: elected State and local officials; law enforcement, civil defense, firefighting, first aid, health, local environmental, hospital, and transportation personnel; broadcast and print media; community groups; and owners and operators of facilities subject to the requirements of this subchapter. (SARA Sec. 301 (c))

Clearly this is an attempt to involve the entire community. It is important to note the inclusion of the media and the potentially critical role they play in disseminating public information in an emergency.

There is no explicit requirement to ameliorate unsafe conditions or to reduce emissions. Clearly the intent here is environmental improvement through informed community pressure. The regulatory approach at best is problematic. It must rely on politically justified minimum standards and spotty monitoring and conformance. A fundamentally different approach is to transcend the entrenched command and control processes and to engage a broader set of actors and stakeholders. Many businesses did not know what regulated materials they had on hand, how much, where, and what this safety concerns were. The mandated inventory is a first step in management and reduction. By requiring firms to communicate their holdings to locals and to develop emergency response plans, additional pressure and insights are focussed on disaster avoidance.

As one might expect, the intent of legislation might be quite different from its actual implementation, particularly a Federal law that mandates a local information-focused activity in perpetuity. Continuing participation after the initial startup and in periods of non-emergency is always a challenge to local groups. While industrial participation nationwide is at the 85% range, neighborhood and environmental group participation is around 50% (Karetz, 1993). There are major regional and local differences in opinion regarding whether LEPCs should play an active or passive role in requiring detailed information from industry, and in publicizing it to the community at large. Almost universally there is a lack of resources to implement this minimally funded mandate.

Some in industry have continued to complain about the complexity and ambiguity of the Title III reporting requirements. Meanwhile, many of the LEPCs are overwhelmed by the volume and format of the hard-copy submittals. It is not uncommon for these submissions to be boxed and placed in dead storage. In contrast, some communities have made considerable progress through the development and integration of databases, LIS, and the integration of EPCRA with emergency 911 systems.

## TOXIC RELEASE INVENTORY

A totally separate part of EPCRA, Section 313, the Toxic Release Inventory (TRI), has played a revolutionary role in empowering community groups and facilitating industry reengineering of chemical usage. TRI is a purely informational policy. All regulated facilities (10 or more employees, SIC Codes 2000-3999, handling threshold or greater amounts of listed chemicals) must annually report to the states on standardized forms quantitative estimates of the release of each chemical to the environment. Releases are grouped by receiving media, air, water, and land. The data is forwarded to the EPA who compile computerized data products. These are readily available on networks, such as the National Library of Medicine's TOXNET, and CD-ROM, along with the MSDS for each chemical. The information is checked and statistically processed by the EPA and disseminated with a simple database program to facilitate access and aggregation. The basic data flow is shown in Figure 8.4.

EPA spatially identifies activities by "facility." A facility consists of:

All buildings, equipment, structures, and other stationary items which are located on a single site or on contiguous or adjacent sites and which are owned or operated by the same person. (EPA, 1991, p. 2)

Because industrial complexes may involve a variety of activities, a facility may be subdivided into separate SIC coded "establishments", with separate forms submitted for each. Facilities are geolocated by latitude and longitude. Additional descriptor requirements include: Dun and Bradstreet Number, EPA RCRA ID Number, NPDES Permit Number(s), Underground Injection Well Code Number as applicable. This reporting requirement provides a means for the Agency to connect key fields in various program-specific databases.

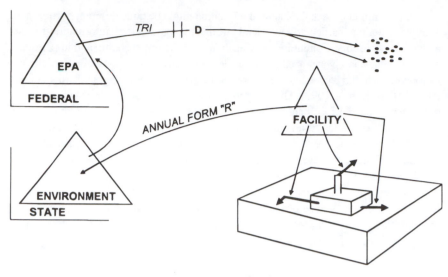

**Figure 8.4. TRI Reporting**

The obvious power of this information requirement is that for the first time it provides an aggregate "look" at many of the releases in a particular area, such as a zip code or municipality. Although the vast majority of these releases are both legal and regulated, the extensive Balkanization of permitting activities reinforced by the proliferation of uncoordinated data systems had prevented governments, communities, and often industry itself from seeing the big picture. Attempting to develop a comprehensive view of a metropolitan area's toxic multi-source regulated loading through FOIA requests was an onerous and often impossible task.

The effect of TRI were immediate and lasting. The press made front page stories of "top ten" polluters, with particular emphasis on local industries. Although most of the releases were legal, pressure quickly mounted on facilites to reduce the use and emissions of regulated substances. The public, which had been poorly informed and inactive through two decades of regulatory programs, could immediately relate to the simple summaries of TRI. Communities and interest groups could use the information to pursue direct negotiations and lawsuits. Many industries responded by developing voluntary, highly publicized programs to reengineer processes so that they exceeded regulatory minimum standards.

Prior to TRI, much of the environmental regulatory process had become a closed dialog between regulators and industry with no external check or balance. At best, companies aimed to meet minimum requirements while maximizing releases.

TRI provides citizens with accurate information about potentially hazardous chemicals and their use so citizens can hold companies accountable... this vast source of data is widely recognized as a powerful force for environmental improvement. (EPA, 1994, p. 1–4)

The availability of the release data provided a new and efficient view of how well the government regulators were implementing their mandates. In addition to mobilizing many communities, the TRI information helped spawn two new movements, environmental justice, and citizen suits against regulatory agencies. Environmental justice is focused on the disproportionate amount of pollution exposure among the poor and minorities. Since the TRI data includes facility zip codes, importing this data into a GIS which includes census data allows spatial analysis of age, sex, race, and income characteristics with toxic releases. Although there are some statistical problems which result from using zip codes rather than geocoding, a strong argument can be made that pollution exposures are neither random nor equally distributed (Monmonier, 1993). The EPA response to the equity issue was discussed in Chapter 5.

Most modern environmental laws provide explicit standing for citizens to sue not just industry for personal damages due to illegal pollution but also governments that fail to enforce the law. A number of NGOs have focused on this important check and balance function using TRI data as a powerful tool for assessing patterns of informal regulatory policies related to industry nonconformance.

Despite the widespread successes of TRI in effecting mandatory and voluntary environmental improvements, there are numerous criticisms of its design and conduct. It was reported that the U.S. General Accounting Office had estimated that, due to SIC classes, chemical lists and threshold quantities, only about 5% of the total chemical emissions actually fell within the mandatory reporting criteria (Moos, 1992, p. 15). The inventory essentially consists of self-reported aggregated data. It has no connection to internal information regarding the industrial unit processes that generated the releases or external systems knowledge building regarding risk assessment dispersion pathways and exposures.

The EPA cautions:

> Total releases or transfer amounts only give part of the story... the number of pounds of chemical releases cannot be directly translated into a quantitative assessment of risk to humans or other organisms. (NY Dept. of Environmental Conservation, 1992, p. iii)

## POLLUTION PREVENTION ACT

The Pollution Prevention Act of 1990 (PPA, 42 *USC* 13101) is an information policy intended to build upon the apparent success of TRI by shifting focus from emission controls to source reduction and elimination:

> As a first step in preventing pollution through source reduction, the EPA must establish a source reduction program which collects and disseminates information. (PPA Sec. 1301(a)(5))

Rather than being designed as a regulation which requires source reduction, the law is based on the presumption that for many facilities source reduction will be a cost-effective investment. Therefore, the basic needs are to make facility owners and operators aware of their current use of toxins, and to educate them regarding successful reduction practices for their industry sector. The first objective is to be met by a major expansion of the TRI reporting. The second objective is the focus of a technical clearinghouse and state assistance programs. In its Source Reduction and Recycling Data Collection section, (Sec. 13106), the Act focuses on the quantity and percent changes of regulated materials in relation to facility production levels. Whereas TRI originally looked solely at outputs ("releases"), PPA goes inside the facility to document source reduction practices, including equipment, redesign, substitution, and improved management.

The implementation of PPA is through a revised "Form R" which expands the concept of "waste" to include "all non-product streams which cross the boundaries of production processes" (Henz, 1992). These reporting elements include both on-site and off-site direct releases and waste stream flows between production, recycling, and waste processing. The basic reported flows are depicted in Figure 8.5.

PPA calls for financial assistance to the states to promote source reduction while placing the data generation burden entirely on the facilities. The intent of this policy appears to be threefold. First, as in TRI, additional public awareness of internal toxic materials handling may lead to additional community pressure on industry to voluntarily reduce volumes. Second, the internal monitoring can be a major self-education step in a facility developing a reengineering program which focuses on cost savings as well as environmental performance. Finally, the specific facility information, as well as the ability to readily aggregate and analyze, enables EPA to analyze current practices. This assists the states in prioritizing sectors and locations for source reduction technical assistance and/or regulatory efforts.

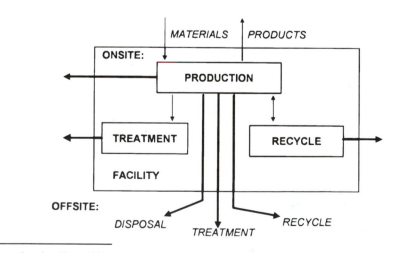

[a] based on Henz, 1992, p. 61

**Figure 8.5. Pollution Prevention Act Reporting[a]**

These potential benefits are offset in part by the protection of "trade secrets," the inherent data quality problems of self-reporting, and the resources and will of the various states and facilities to engage in meaningful source reduction programs.

## FLOOD PLAIN INSURANCE

The evolving role of information as environmental change agent and its relationships to regulatory process is well illustrated in the management of the nation's flood plains. In the early part of our history, floods were considered "acts of God" and their periodic devastation of property and life was accepted. With the industrial revolution, we developed technologies to build dams and levees. A structural era emerged, with agencies such as the Corps of Engineers and Bureau of Reclamation focused on "controlling" floods. Interestingly, the more we spent on structural flood control, the more annual flood damages increased. The structures encouraged flood plain development but were periodically overcome by natural forces.

A fundamental shift in Federal policy occurred in 1968. Congress decided to shift its provision program resources from the construction of projects to the underwriting of property owner insurance, with the passage of National Flood Insurance Act (P.L. 90-44). The focus of this program was to reduce damages in flood plains through land use regulations. Since land use control is a state, not Federal, arena, and since many states delegate this authority to local governments, the federal approach had to be indirect. The program as originally developed by the Federal Emergency Management Agency (FEMA) had four basic components:

A. The Federal government would provide technical and financial assistance for the modelling and mapping (Flood Insurance Rate Maps, FIRM), of flood hazard zones;
B. States would pass flood plain zoning enabling legislation and promulgate model zoning and building codes that local communities could adopt;
C. Communities that adopted flood plain regulations, including the FIRMs as official map, qualified their residents to purchase subsidized flood insurance for existing structures which were retrofitted, and new structures built to standards. Such coverage was previously either unavailable or extremely expensive; and
D. Property owners who meet the regulations are issued building permits and thus qualify for Federally-subsidized flood insurance made available through mortgage lending institutions.

These information components are shown in Figure 8.6.

This ambitious multiparty program, which involves approximately 20,000 municipalities and 2.5 million policies in all 50 states, has had a long history of problems. The fundamental concept of insurance is the viability and size of the risk pool. The goal was to achieve economic self-sufficiency and to eliminate the initial Federal subsidy.

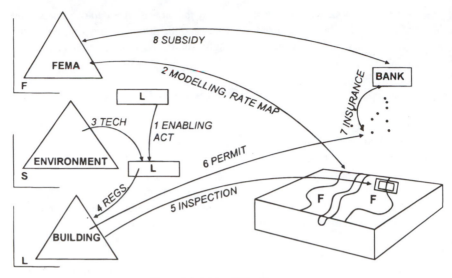

**Figure 8.6.  Federal Flood Insurance**

Low participation rates have led to continuing losses. At first the optional insurance was made mandatory for Federal mortgages. Then the banks were made liable for flood damages that occurred where they failed to require insurance. More recent revisions include the requirement of insurance for all loans, both Federal and private.

The program has had chronic information problems with its maps, its underlying models, and its database. The scales of the FIRM is "approximately" 1" = 2000', with some published maps enlarged to 1" = 1000' and 1" = 500'. The maps depict two basic classes of zones, a central high hazard zone where velocities are rapid, and an adjacent lower hazard area. Rates vary with zonal jurisdiction. While the program insures specific individual structures, and applications require vertical floor elevations to the nearest one-tenth of a foot, FIRMs include no terrain elevation contours and typically do not include structures. These official maps are one or two levels too small for the mandated decision. This scale problem is further exacerbated by the reliance on bank employees to make the jurisdictional decisions. These non-technical staff are advised to make a rather complex set of interpolations solely from the FIRM (Federal Emergency Management Agency, 1995).

The story told to technical people is quite different:

> FIRMs do not include all roads within communities, nor do they depict address, property boundary, or structure location information. As a result, determinations frequently can only be made by using an ancillary source of data, such as a land parcel map, to determine the location of a property on the FIRM. (Federal Emergency Management Agency, 1993, p. v)

The standard paper maps have had a history of distribution problems. The national program involves about 80,000 separate maps, and users request approximately six million copies per year. Backlogged requests and out-of-print sheets are common. In 1989, FEMA started an ambitious program to bypass the map analysis jurisdiction step and replace it with a computer-assisted map address matching directory system. This program failed in large part,

> because the Census Bureau data used to develop the directories contains only 50–60 percent of valid addresses within a community.... Also, about 5 percent of all addresses in the directories have low confidence levels. (Office of the Inspector General, 1992, p. 7)

As discussed in Chapter 6, this problem with address matching was widely known in the cadastral community by the early 1970s.

Following the Inspector General's audit recommendation, FEMA abandoned the directory initiative and has begun a program to provide the paper maps on a CD-ROM as digital FIRMs (DFIRMS) by the year 2000. The format will improve access but will not improve the inherent accuracy of the hard-copy source. Error propagation is expected when the DFIRMs are combined in a GIS with Census TIGER line files for street address matching. The TIGER files use approximate street centerlines and assume street addresses are uniformly spaced along block faces. A "good faith" means of dealing with this location problem is to include a 250' buffer in the DFIRM software. Within this buffer it is recommended that decisions be based on other sources.

A second major map problem is obsolescence. Due to natural geomorphic processes, stream courses and coastlines shift over time. Sometimes this shift is incremental, at other times it happens dramatically as part of a major storm event. As water bodies evolve, their flood zones shift accordingly. FIRMs have a volatile history of obsolescence. This is a double-sided problem. Because they are official maps, when a flood plain shifts, policies and regulations are applied to areas which now may be safe while new hazard areas are ignored. FEMA's program to update maps has been woefully lagging, in large part due to the financial instability of its rate base. Auditors found that from 1984 through 1991, the program had "amended over 10,200 maps," yet the agency "has not published map amendments since 1983" (p. ii). The DFIRMs should permit more efficient currency.

The maps are the products of one-time hydrologic watershed runoff and hydraulic stream corridor geometry modelling. At the time of the creation of the FIRMs, assumptions were made regarding the future development of the watershed. Once the initial predictions were generated, the models were abandoned. The mandated flood plain regulations ignore the watershed as a system, including the importance of controlling runoff from development in the uplands. There is no monitoring of this development. This lack of monitoring and feedback means that even in the absence of major shifts in the stream course, as development deviates from the original long-term assumptions the predicted flood zones will become increasingly inaccurate.

Some geomorphic situations are so complex that the flood mapping program has always been problematic. In western areas with alluvial fans (braided dendritic drainage patterns on fan-like deposits at the base of hills), each time a major flow occurs the stream may follow a different course. Here FEMA has failed to fit a stochastic environment into its deterministic modelling world view. Similarly, areas experiencing flooding combined with coastal erosion do not fit the normal stream flooding scenario.

Ideally, FEMAs insurance policy database would be integrated with an LIS including parcel and structure geolocators that are transactionally updated as field surveys verify policy applications. In reality, the dispersed use of banks as sales agents and the reliance on name-address locators renders the national system essentially incapable of comprehensive spatial analysis. In contrast, many local governments have incorporated flood management in their local LIS.

FEMA's most recent initiative to reduce flood losses goes beyond its foundations of regulation stick and insurance carrot to focus on continuing information and education. Its Community Rating System program has two primary objectives: to increase and maintain flood insurance participation, and to steward the continuous maintenance programs on which the one-time building rehabilitation and new construction permits are premised. The program is optional. Each year a community can elect to participate in variety of flood information and flood damage reduction programs which are above and beyond the required standard development regulations. Examples include: passage of higher regulatory standards than the state minimum, the development of a parcel-based LIS, and the annual maintenance of drainage systems.

Based on the number and type of stewardship activities, the communities get "rated" on a predetermined point basis. By filling out a rating form and submitting documentation of activities to FEMA, municipalities can gain discounts in their flood insurance rates (Federal Emergency Management Agency, 1992). Just like homeowners who install smoke detectors, and drivers with accident-free records, the insurance discount program is intended to reduce damage payouts. It is too early to judge the success of the program.

## RADON

In general, Federal command and control regulations have worked best on major point sources of pollution. Where environmental problems are highly dispersed and/or systems are highly complex, a top-down mechanistic regulatory approach is severely limited. Radon is an example of a highly controversial environmental health problem whose management involves federalism and a significant focus on individual responsibility.

Radon is an invisible, odorless gas that is the natural product of uranium's decay to radium, then radon, then "radon daughters": polonium, lead, and bismuth. The radon daughters are chemically active and may attach to airborne particles which are inhaled. In the lungs, the decay of radon to polonium and polonium to lead both involve the release of ionizing alpha particles which can cause lung cancer. Uranium and radium exist in varying amounts through the earth's crust. Radon gas (Rn222) is chemically inert and can move readily through rock fissures, porous soils, and buildings. It has a half life of approximately four days, while polonium has a half life of three minutes. The U.S. concern about radon emerged at the turn of the century as physicians began to see the correlation between lung cancer in miners and radioactivity in mines. This relation was subsequently confirmed, and the late 1960s the Public Health Service established exposure health standards for miners of 100 picocuries per liter (pCi/l) as a "working level." The cold war arms buildup led to extensive uranium mining, and it was common in mining communities for tailings (a grit-like byproduct of the milling process) to be used as construction fill. The Federal government outlawed this activity in 1966 when it was found that many residences and schools had high radon levels. In 1978, congress passed the Uranium Mill Tailings Radiation Control Act (42 *USC* 7901 et seq.) which provided remediation funds. A level of 4 pCi/l was subsequently established as a the action level.

In 1984 the focus of radon concern dramatically shifted from mining to "natural" occurrences. A nuclear power plant worker in Pennsylvania unintentionally wore his radiation exposure badge home and when he returned to work he set off the monitoring alarms. The company subsequently checked his home and found levels as high as 2,700 pCi/l. The state then surveyed about 3,000 homes in the region and found 40% had levels above 4 pCi/l. This situation had widespread national exposure and it stimulated Congress to authorize research funding.

The radon problem is an excellent example of the multiple roles of information in risk analysis and risk management. The objectives of risk analysis are to identify sources, analyze pathways, quantify exposures, and estimate health outcomes from these exposures. The source identification problem has primarily focussed on regional bedrock geology and soils. In addition to referring to traditional geologic maps, Federal and state agencies have used techniques such as aerial radiometry remote sensing to spatially identify soil radium concentrations.

Although the rock and soil sources may be "natural" sources, building construction and operation (heating, windows,...) are highly localized pathway determinants for human exposure. The measurement of indoor radon levels is highly problematic. Levels vary appreciably daily and seasonally. Many states, such as New York, in an effort to identify which local areas have concentrations of indoor radon, entered into testing agreements with homeowners. The results of these tests are shared with the owner, but by contract are not available to the public. Published analyses depict generalized maps, not specific neighborhoods or buildings.

Two basic measurement techniques are available, short-term (2–7 days), and long-term (4 week to 6 months). The latter has significantly higher precision. Both approaches have major quality control problems in collection and analysis. Collection is easily falsified, as the small collection canister can be readily covered or moved during the exposure period. Canisters are then sent to private laboratories for analysis. The U.S. General Accounting Office conducted a study of these laboratories and found error rates up to 100%. The EPA set up a voluntary Radon Measurement Proficiency Program and hoped that states would formally certify and monitor laboratories. This has generally not occurred (Levins, 1991).

The most controversial aspect in the continuing radon debate is over the setting of standards. The original 4 pCi/l action levels were based on linear interpolations of the much higher miner exposures. In addition, most of the miners smoked. Other studies in the U.S. and Europe had concluded with standards at 10 and 20 pCi/l. Some advocates want the standard lowered to 2 pCi/l. The debate is only moderately informed and highly politicized, as the selection of a standard could affect hundreds of thousands of buildings involving billions of dollars of remediation. After over a decade of study and debate, Congress has not decided to regulate indoor radon levels. This leaves risk management, the action programs designed to reduce the problem, to EPA public information programs, state regulations, and the private sector.

EPA's initial efforts at public education involved Madison Avenue techniques of TV spots, billboards, and mass mailings; their objective was to get homeowners to self-test. The programs have been highly criticized as constituting "scare tactics," with their estimates of cancers and deaths at the extreme high ends of the scientific debate. The results have been ineffective.

If excessive level of indoor radon are found in a building there are a number of structural, mechanical, and operational methods for reducing exposures. These include sealing foundations, installing basement fanned ventilation systems, and modifying user behaviors such as operating doors, windows, and heating. Many buildings are amenable to reductions to acceptable levels for investments of a few thousand dollars. Some would cost much more, and in some situations amelioration is not feasible. Where ventilation systems are installed, their effective operation and maintenance for the life of the building is essential to maintaining occupant safety.

A few states with widespread radon, such as Florida, have adopted regulatory programs. Central Florida includes extensive phosphate deposits which have long been known to include some uranium. In 1986, the state passed the nation's first radon reg-

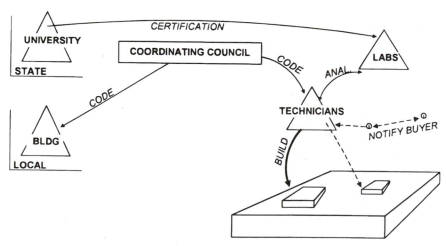

**Figure 8.7.  Florida Radon Program**

ulation, which set 4 pCi/l as the maximum allowable level in new construction. In 1988, this standard was extended to include existing housing (Fla. CS/CS/HB 1420). The 1988 law is multifaceted. It calls for the establishment of a radon control section of the building code for new construction; the certification of radon testing laboratories, measurement and mitigation specialists; development of a trust fund from a tax on new construction; the testing of all schools; and an active public information program (Jurie, 1990).

A critical information component of the Florida approach involves the testing of residences and the notification of buyers. Following EPA recommendations, Florida decided upon a voluntary third party generation and notification approach. Property owners are requested, but not required, to have their homes tested. Information about radon is to be provided in all property sale agreements, with testing at the option of the buyer or seller (Cole, 1993, p. 107). The Florida approach is depicted in Figure 8.7.

## SUMMARY

In Chapter 7, it was shown that a strictly regulatory approach to environmental stewardship has a number of severe limitations. In this chapter, the roles of information policy in supplementing regulations have been introduced. This expansion can engage each of the basic components of deep information.

A typical regulatory approach involves two parties, government and landowner, with knowledge about the environment predetermined in a top down fashion. The regulator either a priori establishes a specification standard, or a performance level for the targeted environmental receiving system. In contrast, deep information calls for broadly shared continuous knowledge building, with a focus on hierarchial environmental processes and the effects of human activities. A simple check on social learning is to examine the actor–stakeholder dataflow diagram. The more parties connected, the more potential for collective knowledge building. The examples of this chapter reveal some of the promise of deep information. Private sector hazardous waste audits, pollution prevention analyses of industrial processes, community inventories of flood prevention practices, and home testing for radon all involve locally responsible parties in situ data collection and assessment.

Environmental knowledge building usually entails the analysis of change over time. Unlike the frequently one-stop regulatory approach, most of these activities are or have the potential to be ongoing. Corporations that develop annual auditing procedures, communities that maintain their flood insurance rating discounts, and homeowners that continue to monitor indoor radon will learn how to reduce environmental loadings. The rich genius loci of the knowledge developed in this manner would be impossible in a top-down specification regulation.

The bottom-up approach is not a panacea. Environmental processes are hierarchial and phenomena manifest at different scales. To build knowledge of regional and larger systems entails the engagement of the second objective of deep information, open plural information generation, analysis, and dissemination. The success of TRI is that the bottom up data are standardized, centralized, analyzed, and redisseminated to communities and interested parties. As discussed, this is not the case in private site audits, or flood insurance. In the former, there is great resistance from the private sector with regard to liability. In the latter, the bureaucracy is still almost entirely focusing on procedural efficiency due to severe budget problems. The use of transactional policy data to update and improve knowledge about the hydrologic flood system has not been engaged, and FEMA is just now recognizing the existence and importance of the natural resources of the riparian zone (Smardon & Felleman, 1995).

The information success of TRI is not a systems knowledge building success. Gross annual estimates of source emissions do not answer basic questions regarding exposure and health. This would require risk assessment modelling of pathways and receiver populations. Similarly, the Florida radon process is an incomplete experiment. Some knowledge will be gained by the monitoring of construction practices. The voluntary third party informing at time of a sale has all the problems of test quality and the lack of a central reporting georeferenced data base. The larger systems question of what should be the exposure standard is beyond the capability of a state to do the long-term scientific research. High quality data on local structures could provide the basis for a large-scale epidemiologic study of long-term; low-level exposures.

The third component of deep information focuses on the downstream information end user, in particular the landowner. The primary means of achieving this record is to create a parcel-based LIS where site specific environmental information is integrated with the land ownership records. Here again, the examples of this chapter show the potential of deep information and the limitations of current practice. For hazardous waste some states now require PIN listings of known waste sites. This minimal connection is a start toward facilitating informed land sales and financing transactions in the light of the major potential liabilities associated with site cleanup. These initiatives are, however, surface first steps, not state LIS approaches. For private site audits, even those publicly reported, there is no such connection. EPA policy, including such programs as TRI, Right-To-Know, and Pollution Prevention, does not legitimize the fundamental role of parcel information. It continues to rely on the administrative convenience of a generalized "facility."

The flood insurance and radon arenas also illustrate the "potential" of LIS rather than the reality. Some municipalities are including these factors in the development of local systems but this is because of the vacuum in state and Federal information. FEMA policies encourage the use of parcel maps but do nothing to encourage the systematic development of state and local LISs, even though misinformation for buyers, sellers, and insurers is probably their number one problem. A simple deed cross-reference that a parcel is in a flood plain would alleviate much of the inefficient "churning" that goes on each time a property is bought, sold, or refinanced.

For radon, the EPA's and Florida's decision to rely on third party voluntary notification of self-test results is a politically convenient but environmentally ignorant resolution. If the bedrock and soil of a site emit radon this is a fundamental characteristic of the real property. If this is a serious matter of public health, then it should be established by standard testing and belongs in the public record of the parcel. Owners and occupants of buildings with mechanical radon ventilation systems need to be periodically notified to test and maintain these systems. Again, a transactionally updated parcel in based record system is the most efficient way to meet this objective.

Regulatory information policy can significantly broaden the scope and effectiveness of traditional regulatory programs. As pointed out in this and the previous chapter, regulatory information is intrinsically ad hoc in time and place. In order to get a more comprehensive view of environmental systems, one must monitor the selected system in a robust fashion. The focus of the next chapter is on science information systems.

## REFERENCES

American Society for Testing Materials. *Environmental Site Audit Transaction Screen Questionnaire.* Philadelphia, PA: The Society, 1993.

Cole, L. *Element of Risk: The Politics of Radon.* New York, Oxford University Press, 1993.

Federal Emergency Management Agency. *How to Use a Flood Map to Determine Flood Risk for a Property.* Washington, D.C.: The Agency, 1995.

Federal Emergency Management Agency. *National Flood Insurance Program Community Rating System Coordinator's Manual.* Washington, D.C.: The Agency, 1992.

Federal Emergency Management Agency. *National Flood Insurance Program: Standards for Digital Flood Insurance Rate Maps.* Washington, D.C.: The Agency, 1993.

Henz, D. "Walk-Through of the New Form R." *Pollution Engineering,* 11 (June 1, 1992): 60–63.

Highland, J. "Can Someone Else Besides a Lawyer Determine Who Really Is Innocent? Technical Perspectives on Due Dilegence and Environmental Assessments," in *Burdens of Environmental Regulation on Private Property Ownership and Business Transactions: Reasonable or Unreasonable?* Washington, D.C.: American Bar Association, 1987, pp. 23–34.

Jacobs, J. *The Death and Life of Great American Cities.* New York: Random House, 1961.

Jurie, J. "Florida Takes Aim at Radon." *Environmental and Urban Issues* (Winter 1990): 1-5.

Karetz, J. Technological Hazards and Their Implication for Natural Hazards Management, *Natural Hazards Management* v. 17 n. 3 (1993) 1:3.

Kass, S. & R. McCarroll. "The Corporate Environmental Audit: Will EPA's New Policy Finally Level the Playing Field?" *Environmental Protection* (June 1995): 45-47.

Levins, H. "Home Safe Home?" *Environmental Protection* (June 1991): 31-38.

Monmonier, M. "Zip Codes, Data Compatibility, and Environmental Racism." *Statistical Computing and Statistical Graphics Nesletter* (Dec. 1993): 14.

Moos, S. "Pollution-Prevention Power to the People," *Pollution Prevention,* 5 (7) (1992):.

NY Dept. of Environmental Conservation. *N.Y. State 1991 Toxic Release Inventory Review.* Albany, NY: The Department, 1992.

Office of the Inspector General, Federal Emergency Management Agency. *Audit of Flood Insurance Mapping Activities.* Washington, D.C.: The Agency, 1992.

Smardon, R. & J. Felleman. *Protecting Flood Plain Resources: A Guidebook for Local Communities.* Washington, D.C.: Federal Interagency Floodplain Management Task Force, 1995.

Turner, S. "Superfund and the Innocent Landowner," in *Impact of Environmental Law on Real Estate and Commercial Transactions*. Albany,NY: New York State Bar Accociation, 1982, pp. 3–10.

U.S. Environmental Protection Agency. *Toxic Release Inventory: 1987-92 CD-ROM User's Manual*. Washington, D.C.: Government Printers Office, 1994.

U.S. Environmental Protection Agency. *TRI Questions and Answers*. Washington, D.C.: The Agency, 1991.

# 9

## Environmental Science Programs

........................

### INTRODUCTION

*I*n Chapter 4, some of the basic characteristics of environmental processes were outlined. These include: complex dynamic systems, hierarchial scales with different phenomena manifesting at different geographic magnitudes, and temporal change, which ranges from cyclical to chaotic. A major factor now affecting most "natural" processes is the output from human activities. As introduced in Section I, a primary objective of deep information is to increase our knowledge of environmental processes and the role of human impacts.

In Chapters 7 and 8 it became clear that regulatory programs, whether focused on direct regulation, or using information as regulation are quite limited as a basis for systems knowledge building. Many regulatory programs are a legislative and agency response to the public's desire to quickly "solve" a problem such as toxic waste or the loss of wetlands. The emphasis is on action not learning. Regulations necessarily focus on a narrow definition of a problem, and are assigned to a narrowly focussed section of an administrative agency. This is a result of both the needs of bureaucracies for reductionism, and the general resistance in our society to any governmental intervention.

From an information perspective, an additional set of factors constrain regulatory programs from effectively building environmental systems knowledge. Regulations are applied in locally site-specific, spatially and temporally ad hoc contexts. Initiation is typically dependent on a proposal for development or change originating from a landowner. Much of the regulatory information is generated by the owner/operator as second party self-reporting with descriptive and predictive information focused on the site and immediate local context. It is common for neither the owner nor the review agency to address those issues related to larger systems scale such as cumulative impacts. This is due to both to resource constraints and the fundamental resistance of transactional decision makers to deal with issues of distributional equity. Finally, there is sparse monitoring of regulated activities. Since all dynamic systems exhibit change over time, the lack of periodic monitoring severs a fundamental feedback which is a prerequisite for learning.

From an historical perspective, these limitations of recent regulatory programs for knowledge building are not necessarily viewed as a fundamental problem. Since the emergence of positivism in the Enlightenment, governments have taken a proactive role in the development of environmental information and knowledge. In activities ranging from the support of individual researchers in laboratories, to the sponsorship of exploratory parties, government science programs have played a major role in learning about our environment for centuries.

Environmental science programs each have a particular set of descriptive or learning objectives. They are designed by specialized professionals to generate information that meets explicit standards of quality using methods of sampling and statistics. Much of this activity has traditionally focussed on descriptive classification and mapping of environmental features such as terrain. Spatially and temporally these programs have widely different structure. Some, such as the soil survey, have complete areal coverage, and are systematically conducted county by county for the entire country with a cycle of once every 40 or 50 years. In contrast, the network of stream gauges includes hundreds of selected point sampling locations with continuous flow monitoring.

The outputs of these science programs may serve all major sectors of the actor–stakeholder diagram. One traditional audience is policy input to the legislature. Examples range from the use of the census of population for reapportionment, to the use of Lewis and Clark's explorations to develop settlement programs, to the ongoing Natural Resources Inventory (NRI) discussed below. Another end user group has traditionally been business. The Department of Agriculture, with its support of land grant colleges, experiment stations, and cooperative extension programs, focusses much of its science on information and technology transfer to the private sector. Similarly, business is a primary user of the Federally-generated weather forecasts.

Governmental agencies also make extensive use of the science programs in the development and implementation of their other functions. As discussed in previous chapters, the USGS quadrangle maps have become the de facto base maps for many regulatory programs. Frequently, the government's science programs have generated a highly diverse web of downstream end users, including business, NGOs, and educational institutions. All constituencies play a vital role in generating the continued political support necessary to maintain public funding.

For the United States, with its history of geographic expansion and settlement which has occurred throughout the agricultural, industrial, and post-industrial eras, the role of environmental knowledge building science programs has long been institutionalized in our Federal and state governments. In this chapter, we will examine some selected non-military science programs from the perspectives of information policy and deep information. (For an excellent analysis of the additional information filters involved with national security related issues, the reader is referred to McClure & Hernon, 1989).

## THE CENSUS OF POPULATION AND HOUSING

The census of population is the oldest and, from an information policy perspective, the most robust of our major environmental science programs. The constitution calls for a decennial census of population, which was a critical component of the political resolution of representation in a rapidly growing country. Starting in 1790 as a ad hoc enumeration process not assigned to a bureaucratic agency, the scope and rigor of the process rapidly evolved. By the 1840s the American Statistical Association began providing critiques of the census methods, and for the 1850 count the census function was assigned to the Department of Interior. In 1902, a permanent Bureau of the Census was established. The Bureau has been a world leader in the development of survey and analysis techniques, and in recent decades the development of large-scale computing and arc-node GIS systems (Clemence, 1985).

The census provides a unique longitudinal view of the composition, spatial location, housing stock, and transportation of the nation's residents. The primary vehicle for data collection is the household questionnaire, which is implemented either as an interview by a trained enumerator, or increasingly as a second party self-reporting mail survey. There are two sets of questionnaires, the "short" and "long" forms. The former focuses on household composition, sex, age, race, marital status, origin, and housing type, improvements (indoor plumbing), size, and cost. The short form questions are administered universally. A sample of residences get the long form, which has over forty additional questions ranging from education, employment, and transportation to work, crop sales, and sewage disposal. From these direct questions, a series of "derived" variables can be generated, such as population density and farm residence.

From early in the nineteenth century, the public and private sectors have viewed the census as a vehicle for satisfying the data needs of a broad spectrum of end users. These pressures have grown geometrically as governmental social programs have mandated the use of census statistics, and the manufacturing, marketing, and service sectors have become increasingly demographic sensitive. The selection of questions for inclusion and the design of dissemination products are critical policy decisions.

The Census Bureau maintains an extensive advisory system of representatives from the Federal, state, and local governments, and private information industry. For the 1990 census, the Bureau and the OMB formed a Federal agency coordinating group to advise on national data needs. Committees were also formed to address the needs of distinct minority groups. The Bureau maintained standing advisory groups representing the major end-user professional organizations and business associations.

The question selection process is severely constrained. The number of questions is basically fixed, due to the rapid fall off of response to questionnaires that are too long. Therefore, a new question must supersede a previous one. This has major implications for the analysis of trends. As discussed below, the census is spatially keyed to small areas such as the block. Data needs at this scale have priority, as do questions which generate data mandated by Federal programs. Once the topics have been selected, extensive quality control testing is done for question wording and sample design.

Scale hierarchy is the fundamental principle of census geography. Although the original constitutional mandate simply called for population totals by state to apportion representatives, because the survey of households is intrinsically site-specific the ability to generate local and regional aggregates has been the basic strength of the census program. The hierarchy includes: the nation; four regions ("northeast"), divisions ("middle atlantic"); states; counties and equivalents; county subdivisions (such as towns); places (inhabited clusters which may cross government bounds); census tracts and block numbering areas (contain 2500-8000 inhabitants in relatively homogeneous socio-economic conditions); block groups; and blocks (Marx, 1990). The hierarchy is shown in Figure 9.1 (Bureau of Census, 1992).

The complexity of the sub-state levels of this hierarchy is a reflection of both the spatial diversity of development patterns in the country, and the varied historical legacies of local jurisdictional subdivision. The Census Bureau relies on the states to provide updated maps of changes in counties and minor civil divisions such as cities. State and local end users play a critical role in the delineation of tracts and blocks, and in the updating of base maps, including street names and address ranges. Prior to 1960, the Bureau relied on maps provided by the states and local governments. Many of these maps were 10, 20, or more years old and had led to serious quality and efficiency problems. With the advent of large-scale computing, the Bureau undertook a leadership role in the development of nationally standardized digital base mapping which had evolved by the 1990 census into TIGER (Topologically Integrated Geographic Encoding and Referencing system).

TIGER is based on USGS 1:100,000 scale maps which were derived from mosaics of the 1:24,000 scale quads. Roads and development depicted on the maps were updated from recent aerial photographs. The TIGER system is an arc-node GIS as shown in Figure 4.9, with data layers including roads, hydrography, and railroads from the USGS and the census geography depicted in Figure 9.1. The road attributes

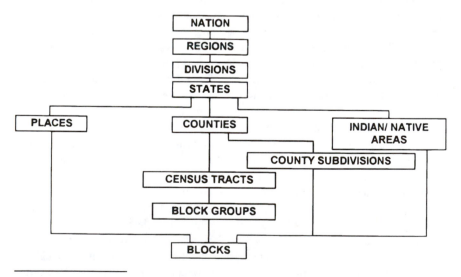

[a] based on Bureau of Census, 1992, p. 3

**Figure 9.1. Census Spatial Hierarchy**

include address ranges, and additional records include zip code boundaries. This structure plays a critical role in the design of the spatial sampling, the implementation of the door-to-door and mail surveys, and in subsequent statistical analysis and dissemination (Carbaugh & Marx, 1990).

The census has thousands of end users, ranging from Congress, to administrative agencies, to marketers and environmental analysts. In order to use its scarce resources effectively and to utilize the energies of the private sector, the Bureau focuses on data analysis and the timely release of a series of standard digital data products. Commercial firms produce a host of value-added products. Throughout the history of the census a critical issue has been protecting the confidentiality of the responders. Over 100 years before the Privacy Act, the confidentiality of citizen responses was debated. Starting in 1850, the Federal government eliminated the required local display of census responses and shifted to release only aggregate numbers. Concerns included misuse of the data and growing resistance among the public to answering the questions. However, copies of the raw data were provided to county courts. This data was also misused, and beginning in 1880 confidentiality was controlled by legislation.

The Bureau has developed elaborate confidentiality procedures for handling data, ranging from swearing-in census takers, to using sealed envelopes (rather than cheaper post cards), and stripping names and addresses from processed records. By law, the Census cannot publish data that can be associated with an individual. By 1940, the Bureau had become sensitive to summary data where categories had few respondents, and by 1960 statistical methods were put in place to address this problem systematically (Courtland, 1985).

For environmental analysis, the Census plays a number of important roles. Although the main function of the effort is enumeration, the Bureau has developed sophisticated modelling approaches to forecasting future populations based on vital (birth and death) statistics, in and out migration, and associated economic forecasts. These forecasts play a central role in project development, such as transportation planning and impact analysis. A growing component of impact analysis involves risk assessment. Census data are essential in evaluating plume exposures and in studying the controversial issue of environmental equity. The TIGER product, which was developed as a means to a more efficient and effective census, has in many ways become an end in itself. TIGER line files and associated attribute tables are the basis for many local low-cost GIS systems. The private sector has aggressively entered the market to update and spatially enhance the TIGER system. The importance of this arena has come to the forefront in the effort by the Bureau to enlist state, local, and private participation in cooperatively updating the digital files for the 2000 census (Sperling, 1995).

## COORDINATION OF WETLAND INVENTORIES

The census provides an excellent example of a continuing longitudinal learning process that has evolved to incorporate both new information technologies and a highly participatory form of study design, data generation, and dissemination. The census originated as a narrow single-purpose activity, and became more comprehensive in scope as the public and private sectors realized the information potentials of an expanded survey.

A fundamentally different challenge to environmental systems learning was seen in the Greenetowne example, where agency Balkanization created fractured, dispersed sets of environmental information. A major example of this disjointed approach at the Federal level has been wetland inventories. With the rise of our understanding of the role of wetlands in natural systems, an array of study programs and protective regulations were legislated. For over a decade, these regulatory programs have been highly politicized, with opponents raising questions regarding property rights and widespread inconsistencies in jurisdictional delineations.

In Chapter 7, we reviewed the official map difficulties of New York's wetland regulatory program. That discussion was focussed on the state law, and did not deal in depth with the complications that wetlands are also regulated by the Corps of Engineers, who use a different classification system and a different size threshold. At the Federal level this problem increases geometrically as five major agencies, The Corps, the Fish and Wildlife Service (FWS), the Natural Resources Conservation Service (NCRS), EPA, and NOAA all have separate wetland mandates.

Although the primary political outcry is over regulations, the more fundamental issue identified by the National Research Council is the need for coordinated information policy. Since the agencies use different classifications, have different mapping schedules and different spatial jurisdictions, it is not possible for the administration or Congress to get a comprehensive longitudinal view of the nation's wetland resources. In 1990, the president formed the Domestic Policy Council—Interagency Wetlands Task Force to review current inventories, assess needs, and provide coordination recommendations (Mapping Sciences Committee, 1993).

In 1993, the White House Office on Environmental Policy announced:

> the administration will direct... the Wetlands Subcommittee of the FGDC to complete reconciliation and integration of all Federal agency wetland inventory activities. (Shapiro, 1995, p. ix)

This stated policy is both symbolic and substantive. The FGDC is not well constituted to engage the complex issues of the use of inventories in regulations and provision programs. Rather, as the title of the Shapiro report states, it is focussed on the more scientific concern of inventories for the analysis of status and trends. The following discussion is based primarily on Shapiro.

The FGDC's response has been to develop a four-stage program:

Task I— coordination of terms, definitions, and classifications. A basic consensus has been reached among FWS, NCRS, NOAA, EPA, and the USGS to use an ecologically based hierarchial classification approach.

Task II— coordination of data collection processes and reports. The response here has two thrusts. First, it is proposed that future reports include a "crosswalk" that will explain the relationships between the data sets used by the reporting agency and those generated by other agencies. The second thrust is to share remote sensing efforts and disseminate data through a single digital system.

Task III—analyze the consistency of agency data (discussed below).

Task IV—recommend policy changes.

The Task III effort entailed a case study of complex forested and open wetlands which were mapped independently using seven different Federal approaches and the method used by the state of Maryland. Each analysis was entered as an arc-node GIS layer so that comparisons could be overlayed. In addition to varying classification systems, the approaches used differing scales and types of remote sensing and different levels of field checks. Most produced thematic maps, such as the Fish and Wildlife Service's series depicted in Figure 9.2a, but one utilized point samples.

The existing Balkanization is a clear reflection of differing enabling purposes. The FWS's National Wetlands Inventory is mandated by the Emergency Wetlands Resources Act of 1986 (16 *USC* 3901 et seq.). It required the thematic mapping of all wetlands in the contiguous states by 1998 and was amended subsequently to include Alaska, and to convert to a digital process by 2004.

The NRCS has two different wetland inventory programs. Wetlands are but one of many categories included in the National Resources Inventory (NRI) point-based sample discussed below. In addition, the Food Security Act of 1985 (16 *USC* 3801 et seq.) included the "Swamp Buster" provisions which eliminated the use of agricultural supports to drain wetlands. The agency delineates wetlands using a combination of soil maps, photo interpretation, and field work. In 1993, the administration designated the NRCS wetland inventory as the "final" government position on Swampbuster and Clean Water Act designations affecting agricultural lands. The EPA and the Corps, who have joint jurisdiction over the Clean Water Act's regulation of wetlands, subsequently signed agreement memoranda with NRCS implementing this policy.

Under the Magnuson Fishery Conservation and Management Act of 1976 (16 *USC* 2001 et seq.), NOAA is charged with monitoring and protecting fish habitats, which include coastal wetlands. In 1990, NOAA started the Coastal Change Analysis Program, which takes a comprehensive look at both wetlands and associated upland subsystems. The program is designed to monitor coasts nationally on a five-year cycle, with more frequent analysis of areas affected by coastal storms and oil spills.

The Task III comparative study also included inventories developed by EPA's Environmental Monitoring and Assessment Program (EMAP), discussed below, and the USGS. In the comparison it was not possible to directly evaluate accuracy,

> because there is no standard of correct wetland classification with which to compare the various data sets.(Shapiro, 1995 p. xi)

Therefore, the study focused on precision (consistency).

The results are not encouraging. The four polygon data sets disagree on over 90% of the area that at least one approach classified as wetland. Since the NRCS wetland inventory approach is used in a service/regulatory program, it appeared to conservatively overestimate, but when it was removed from the comparison there was still an 80% disagreement. In comparing the polygon data with the NRI's point sample approach, there was a 99% disagreement.

The FGDC is struggling to deal with this issue. As discussed in Chapter 4, and portrayed in Figure 4.6, the traditional mechanical hard-edged thematic mapping technique used by all the agencies is fundamentally unsuitable for fine-grained, dynamic phenomena such as wetlands. The Committee is considering the need for a new hybrid category, mixed wetland–upland. This is a modest recognition of the critical role that edges play in the environment, as depicted in Figure 4.7.

In an important corollary effort not assessed by the FGDC, the state of Maryland, which has a proactive wetland regulation program, has created an arc-node GIS capability which includes digital orthophotos, wetland delineations, and property parcel bounds. This system can overcome much of the misinformation institutionalized in New York's program (Maryland, 1995). To date the FGDC Wetlands subcommittee has tended to ignore the regulatory program information policy dimensions of the White House policy. This is a particularly narrow view because, if the reconciled integrated national wetland information is ever to meaningfully inform policy makers, the first questions to be asked is: how many landowners and what value of land is to be affected by the policy? Only the NRI approach discussed below integrates both resource and some cadastral data.

## NATIONAL RESOURCES INVENTORY

One of the inventory programs assessed in the comparative wetlands study is the USDA's National Resources Inventory (NRI). The agency has been classifying and mapping soils for a century. The primary focus of the soil surveys has been to provide information to landowners, originally farmers and in recent decades developers and planners. For over 60 years the USDA Natural Resources Conservation Service (NRCS, formerly the Soil Conservation Service (SCS)) has inventoried resources with the objective of informing national policy, starting with the dust bowl initiated National Erosion Reconnaissance Survey in 1934. This field effort led to the passage of the Soil Conservation Act of 1935. In 1945, the SCS published *Soil and Water Conservation Needs Estimates for the U.S.,* which was based on multiple secondary sources. In 1958, a National Inventory of Soil and Water Conservation Needs was established by Secretary of the Department of Agriculture Memorandum No. 1396. It stated that:

(the Inventory) will be made and kept current.... From it the Department could arrive at reasonable estimates of the magnitude and urgency of the various conservation measures needed to maintain and improve the country's productive capacity for all the people. (Harlow, 1994, p. 2)

The 1958 Inventory utilized a statistical sampling design. Stratified by counties, random areas representing 2% samples were generated with field mapping of soils and land cover of non-urban sample areas. In 1967, the Inventory was updated. Changes included the addition of random point samples within the area samples.

With the coming of the modern environmental movement as represented by NEPA in 1970, and the pervasive concern over the loss of resources due to interstate highways and suburban sprawl, the focus of many traditional single-purpose Federal agencies was broadened. The Rural Development Act of 1972 (P.L. 92-419 Sec. 302) mandated the Secretary of Agriculture to issue a "land inventory report" every five years. In addition to soil and water conservation, objectives included "balanced rural and urban growth," "flood plains," and "degradation to the environment." To implement this law the Department established a small, full-time, budgeted inventory unit. In 1977, the continuing NRI was begun.

Through legislative changes and the utilization of computer advances, the NRI has rapidly evolved. For example, the Soil and Water Conservation Act of 1977 (16 *USC* 2001 et seq.) required appraisal of fish and wildlife habitats, as well as the "use of data by other Federal, State, and local agencies and organizations." While mandates continued broadening the inventory's scope, the agency has been under continuous budget and staffing pressures due in large part to the in situ staffing of data collection. A move to remote sensing and GIS has partially resolved the staffing difficulties. By 1992, the Department was able for the first time to produce statistically valid ten year trend analyses (Harlow, 1994).

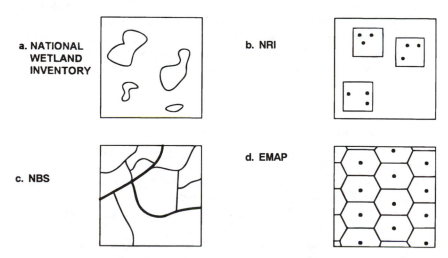

**Figure 9.2. Science Program Spatial Samples**

The NRI database consists of three primary files: county, soils, and sample. The county file includes census and basic descriptive information about individual counties, such as their area, surfacewater, transportation, and Federal land ownership. The second data set is the national file of attributes for mapped soils. The third data file includes attributes for the sample locations which consists of a set of square areal "units" ranging in size from 40 to 640 acres depending on region, and points. Hundreds of potential values may be recorded for each sample location, with the majority focussed on land cover and agricultural practices. Attribute values were determined by photo interpretation, field samples, and analysis of USGS and soil maps and departmental Conservation Reserve program files. There were over 300,000 areal sampling units, each containing three random point samples, in the 1992 NRI. The design is depicted in Figure 9.2b.

The NRI is designed to produce aggregate shallow temporal knowledge for non-Federal lands. Statistical descriptive trends, such as regional decreases in wetlands, can be generated. For some topics, these may be correlated with major shifts in Federal agricultural programs such as Swamp Buster discussed above. This information complements the traditional econometric modelling which is the primary policy analysis tool of the Agriculture Department. The 1992 NRI was processed by 1993 and used in the negotiations of the 1995 farm bill. Since land use change is also strongly influenced by state and local policies and local economies, the Federal policy-natural resource change relationships are intrinsically correlative.

Although the NRI provides a high-quality point sample database of hundreds of attributes per data point, it was never intended to be used in environmental systems modelling. This is due to three fundamental NRI characteristics: scale, randomness, and confidentiality. Since data collection is expensive, the NRI's stratified/random sample was carefully designed to optimize cost-effectiveness by minimizing the number of sample points needed to satisfy national policy needs. In many cases there may not be enough data points to provide meaningful information regarding local or subregional environmental analysis needs. For example, although counties were explicitly part of the design, multi-county "Major Land Resource Areas" are probably the smallest consistent aggregation scale. New York state has ten such areas, three internal and eight shared with neighboring states.

For environmental process modelling, the major movement over the past two decades has been a shift from "lumped" to "distributed" parameters (see discussion in Chapter 4). The NRI's random points are of little use in distributed models. For example, the Department of Agriculture's own computer modelling of non-point pollution in the Lake Champlain Basin, which is only part of the multistate St. Lawrence/Champlain Major Land Resource Area, used its GIS system with thematic mapped aerial coverage of local tributary watersheds.

Even if the models could make use of the random data through contouring, the data is not available outside the agency due to privacy and confidentiality. Privacy is important in gaining landowner access (just as it was in maintaining census participation). The sample certainly reveals that some farmers are not conforming with their signed land management requirements of programs such as Swamp Buster. It is essential that this data be acquired and the "price" is that it will not be used for regulatory purposes. A critical temporal design component of the NRI is that each analysis includes some new locations and some repeat locations. In order to protect the reliability of the repeat location sites, confidentiality policy was developed.

The NRI developed as an internal departmental program with the sole objective of informing national agricultural and environmental policy. As Federal budgets have decreased and increasing emphasis has been placed on "performance", the NRI has come under increasing scrutiny. As early as the 1982 inventory, the Department had inadequate resources to meet the increased Congressional reporting mandates and a number of data "gaps" were generated which later inventories closed by interpolation estimates. In a 1985 analysis, OMB recommended that the inventory be shifted from a five- to a ten-year cycle, but this was inconsistent with previous Congressional reporting requirements that called for five-year cycles.

USDA has aggressively adopted information technology to increase the efficiency of the inventory and the effectiveness of dissemination. The next inventory will make major increases in the utilization of remote sensing and GIS to generate and analyze attribute data. To improve internal utilization of the NRI information and broaden the network of downstream end users, the 1992 inventory process included the development of a user-friendly database program and the release of the information in digital form. An important feature of the program is the error trapping of regional and local queries that exceed the statistical validity of the sampling design. Many potential users such as impact analysts want local information which is beyond the resolution of the NRI. Another important information policy feature of the dissemination effort is the maintenance of precise sample location confidentiality.

Except for some state agricultural programs, which have worked cooperatively with USDA to increase the sample size to provide more detailed and useful localized information, the Department has not been able to generate a significant external constituency for this major undertaking. The program faces a future of continued resource reductions. One proposal for dramatically changing the implementation of the inventory is to shift it from an episodic process to a "continuing" inventory. This would both shift the funding from a periodic, controversial add-on to a standard internal item, and level out the training, site analysis, and data processing activity loadings. The potential is to transform the NRI from a free-standing information activity to a more integrated corporate data model approach (USDA, 1990).

## NATIONAL BIOLOGICAL SURVEY

A key element of the wetland information coordination problem discussed above was the Balkanization of classification standards and mapping approaches. Although wetlands have been at the center of attention due to regulatory and "takings" issues, they represent only a small fraction of a much deeper problem facing environmental sustainability. The United States lacks standard descriptive information about the types and spatial extent of our rich biodiversity. The USGS base maps just differentiate "forest cover." Most regional and local studies traditionally have focused on a resource attribute such as potential timber production, and were designed assuming the constraints of single hard-copy thematic maps. These traditional management-based approaches are unsuitable for analyses concerned with biodiversity and sustainability.

Leadership in concern for biodiversity has come in the past two decades not from the public sector, but rather from NGOs such as Audubon, the Sierra Club, Ducks Unlimited, and in particular the Nature Conservancy. Underlying the latter's long-term strategy of removing critical habitat from the threat of development, it has developed a national network of Natural Heritage Programs in cooperation with the states to classify and map unique resources. The classifications, based largely on site work at the plot and community scales, are managed in a database. The database facilitates ranking related to uniqueness.

The development of a standard national approach to vegetative classification and mapping has traditionally not been by a formal policy mandate. As more and more Federal agencies have moved to include ecosystem management and sustainable development in their mission statements, the need for a comprehensive understanding of the resource base has become apparent.

> An ecosystem is broadly defined as a unit of the landscape that is somehow 'tied together' through a shared set of ecological processes. These ecosystems may be delineated using different ecological variables at multiple scales. At this point, different agencies are delineating ecosystems units that will help them address their agency-specific objectives. Variation in the definitions between agencies makes it important to apply common descriptions to these units across the physical and administrative landscape. (Nature Conservancy, 1994, p. 2).

Although Congress had not mandated it due to property rights resistance, the need for a unified comprehensive approach became so overwhelming that finally, in 1993, the Secretary of the Interior internally reorganized the Department to create the National Biological Survey (NBS). NBS' two main objectives are to collect and analyze data and to disseminate information regarding the nation's biological resources. As part of the emerging National Spatial Data Infrastructure, NBS is to provide a National Biological Information Infrastructure.

The first major initiative of NBS related to development of standardized classification is a joint project with the National Park Service to generate classifications of all of the nation's parks. This program is strategic from three perspectives. The parks include a vast spectrum of regional and local diversity, which will provide a robust test for the development of a national classification system. National parks are high profile, with considerable public attention both to their unique resources, and the increasing stresses of overuse and declining funding. Finally, the parks are Federal property managed by the Interior Department. This is critical because surveys include extensive ground checking and the project involves both government and Nature Conservancy staff in a joint venture. The property rights opponents are extremely opposed to entry onto private lands, particularly if the result may be the finding of an endangered species.

An underlying assumption in the process is that vegetation is an ecological integrator that synthesizes geology and soils, climate, and local management practices. In this capacity it can indicate both the capacity to support wildlife and to display the results of stresses such as soil erosion and pollution loadings.

The Standardized National Vegetation Classification System was developed by the Conservancy and the NBS in coordination with the Fish and Wildlife Service, the Forest Service, the Park Service, EPA, and a number of universities. It fully utilizes the concepts of systems and hierarchies and the information technologies of remote sensing, GIS, and database.

The new national system is based on existing vegetation, including natural, seminatural, modified, and cultural. It utilizes a hybrid top-down physiognomic and bottom-up floristic approach. The former lends itself to remote sensing and GIS analysis, while the latter involves field identification. As the block was the basic unit in the census, the "community element" is the basis of this system. These elements represent:

> structurally and floristically homogeneous stands of vegetation that occur in a relatively uniform environmental setting. (p. 5)

Again paralleling the census, the national system involves a seven level hierarchy: system, physiognomic class, physiognomic subclass, formation group, formation, alliance, and community element. Different scales have importance for different ecosystem functions. One key product will be arc-node GIS thematic map layers, as shown in Figure 9.2c.

As the Phase III comparative study of the wetlands coordination project revealed, different mapping approaches yield different results. Although the NBS' classification system in many ways parallels a number of existing Federal, state, and nonprofit efforts, its implementation for the national parks and ultimately for the country entails an entirely new analysis. In addition to a uniform classification, this is a result of the NBS' rigorous quality control system, and its integrated design of GIS and database structures.

Although this comprehensive approach has the broad support of the technical community and environmental organizations, its implementation is in doubt. Budget cuts have shifted the park deadlines. The NBS program has been moved to the Geological Survey and does not have congressional support. Extension of the NBS to private lands remains problematic from both funding and site access perspectives. Finally, efforts to date have focused on developing the system and initially implementing it in selected locations. No policy has been developed regarding temporal monitoring and updating.

## ENVIRONMENTAL MONITORING AND
## ASSESSMENT PROGRAM

Each of the above science program efforts are designed to nationally assess a set of environmentally significant resources, but the scope of each is necessarily limited due to agency priorities and resource constraints. The most ambitious environmental science program to date is EPA's EMAP. Throughout the 1970s and 1980s, as EPA multiplied its regulatory programs due to legislative mandates, criticism rose that it was unclear that the regulations were in fact protecting the environment. The EPA's origins were deeply rooted in pollution prevention related to human health. In regard to ecosystem sustainability, such as the role of acid rain in the Adirondacks, there was frequently little systems knowledge on which to base the regulatory standards. The criticism was widespread and included both industry and the environmental community.

In 1988, EPA's Science Advisory Board concluded that EPA should develop the internal capacity to move from a crisis-oriented reactionary mode that had dominated its history, to a proactive anticipatory mode. This would entail both a major deepening of its ecosystems understanding, and a comprehensive broadening of its varied chemical focused monitoring programs. The Board recommended that the EPA:

> conduct a regular and systematic monitoring assessment of the status of the American environment, applying this knowledge to determine the status of representative ecological systems, as well as reaching conclusions on regional and local scales. (Committee to Review EPA's EMAP, 1992, p. 3)

In response to the Board, EPA created EMAP in 1990, and charged it with four ambitious objectives:

1. estimate the current status, trends, and changes in newly-developed indicators of the conditions of the Nation's ecological resources with known confidence;
2. estimate the distribution and extent of the Nation's ecological resources;
3. seek associations between selected indicators of natural and anthropogenic stresses and indicators of the condition of ecological resources; and
4. provide annual statistical summaries and periodic assessments of the Nation's ecological resources. (Martinko, 1993, p. 1)

There are a number of strategic information policy elements in this objective set. First, EPA is unilaterally declaring itself the primary Federal agency for ecological information, an arena in which it had little history. Second, the program's success rests entirely on the identification and continuous monitoring of "newly-developed indicators," a fundamental shift from traditional chemical monitoring, such as dissolved oxygen, to include biomonitoring. Finally, the focus was on policy development, particularly the relationships between human activities and ecosystem health.

Since its inception, EMAP has been controversial in the technical community, in other Federal agencies, and in Congress. One of the original areas of concern was over its proposal for spatial sampling. While the FWS' wetland mapping and NBS' vegetation inventories adopted uniform thematic mapping as the foundation for science and program information, the NRI had opted for spatial sampling to efficiently assess trends. The problem is particularly critical in the environment arena with its hierarchial spatial scales. The EMAP resolution was to use GIS to establish a uniformly distributed hierarchial set of hexagons which could subsequently be sampled at larger scales.

The originally proposed design includes the hexagonal tessellation incorporating two nested hierarchial scales. "Tier 1" covers the U.S. with 12,600 evenly spaced hexagons 40 sq. km. each in area. The center of each hexagon represents a sample point. This is shown in Figure 9.2d. From this regular sample estimates can be derived of areal extent and numbers of entities such as water bodies. "Tier 2" was intended to report regional status and trends. Statistical sampling would be used, and the spatial scale of sampling would be increased to reflect the scarcity of the resource to be analyzed. Tier 2 sampling would rotate on a four-year cycle.

This initial sampling design was done without the knowledge gained from application to see if it would adequately provide the required data. In addition to the need for testing, a National Research Council review committee concluded:

> The committee also emphasizes that Tier I must be done *completely* to be useful. Enough money must be budgeted to ensure complete coverage of Tier I because absence of data will compromise Tier I's usefulness. (Committee to Review EPA's EMAP, 1992, p. 6)

The committee also was concerned about the potential bias in the data resulting from landowners refusing site access to inventory staff.

To date, the EMAP program has been in a pilot stage focused on the analysis of selected ecosystem types in pursuit of both testing the sampling design and identifying the "indicators." Landscapes studied include: agriculture, arid, estuaries, forests, Great Lakes, surface waters, and wetlands. For the Chesapeake, a 70 sq. km. hexagonal grid was utilized in the main embayment. For the adjoining tidal rivers, a centerline with 25 km. station points was estimated and a random sample was generated along the centerline. All of these points were included in a four-year sample rotation (Copeland et al., 1994).

In reviewing the Lake pilot project the National Research Council concluded that the Tier 2 sampling design with stratified large and small lakes appeared adequate. However, the pilot was found to be severely limited because it did not include the lakes' watersheds. This omission undermines the program's basic objective of linking ecological conditions to environmental stresses. In addition, a fundamental design criticism was that a standardized temporal sampling approach is insensitive to significant short-term changes in many environmental processes (Committee to Review EPA's EMAP, 1994).

Although the changes in the sampling design were to be expected in the pilots, the articulation of indicators, which is the key to the entire enterprise, is more problematic. The original 1992 review questioned the use of a peer review approach to indicator development. The 1994 NRC review concluded that a peer review approach which relies on numerical indices is overly simplistic. It criticized the program for systematically failing in the past to provide program-wide management for indicator development.

These problems are symptomatic of an overambitious program initiative coinciding with a fundamental downsizing of the Federal government. Although EPA had a rational bureaucratic and professional basis for initiating EMAP, it never had either Congressional support or that of significant external constituencies, except the state and university communities, which were funded for pilot projects. A fundamental flaw was the lack of serious design coordination with other Federal agencies. In the summer of 1995, the EPA drastically scaled down its program focus, abandoning its national scope and prioritizing three environmentally high profile regions: the Pacific Northwest, the Everglades, and Mid-Atlantic coastal watershed. The search for operational indicators will be the main agenda (Stone, 1995).

The program's scientific reviewers are skeptical that the program will ever meet its original objectives. In analyzing its information system, they concluded that traditional relational databases are not adequate for scientific endeavors of this scale. It found EPA's management information systems approach to the challenge insufficient to meet the diverse needs of distributed end users (Committee to Review, 1995).

By 1996, the EPA appeared to have abandoned the objectives of EMAP, and the central role of the agency in ecosystem science. In establishing a strategic set of performance goals for the environment, the agency identifies "healthy terrestrial ecosystems" as one of 12 major action arenas. After symbolically recognizing the importance of ecosystems and biodiversity to the nation's long-term health, and the "patchwork" of current Federal activities, the EPA proposes two major strategic shifts. First, the activity of improving the scientific knowledge base through continual monitoring is explicitly associated with the other resource agencies such as the National Biological Service and agriculture's NRI, as well as the NGO work of the Nature Conservancy. EMAP is not mentioned.

Second, the previous study design, which included a linked hierarchy of system scales from local to regional to national, is abandoned. In its place is EPA's "new approach," "community-based environmental protection" (EPA, 1996).

## SUMMARY

With the intrinsic limitations of administrative regulatory and resource transfer programs to build environmental systems knowledge discussed in Chapter 8, attention was turned in this chapter to the promise of the public scientific enterprise. Any such endeavor must take place within the reality check of Karplus' model, depicted in Figure 1.5, which warns that the dynamic complexity of natural systems is only partially predictable. From the perspectives of deep information the various programs discussed above have potentials and problems.

The deep information objective component of knowledge building has three elements, shared knowledge building, hierarchial environmental processes, and the linkage between natural processes and human activities. Initially, all of the programs examined, with the exception of the joint effort of the National Biological Survey, the National Park Service, and the Nature Conservancy, began as individual agency undertakings. Jurisdictional and resource limits have dramatically shifted the current climate to one of science and monitoring partnerships. This poses significant challenges to efforts such as the NRI and EMAP, whose underlying statistical design were not developed in a partnership context.

The second deep information component is openness in information generation, analysis, and dissemination. Here the rapid revolution of the FGDC and its associated metadata and content standards, which are implementations of the Federal dissemination policies discussed in Chapter 5, have exhibited considerable progress in dissemination. The widespread utilization of GIS and associated spatial and environmental models has led to major advances in shared analysis. Data generation, however, poses a dilemma. Shared utilization of remote sensing has increased generation, but resource constraints and limits on private property access severely limit critical on-site data collection.

The final component of deep information emphasizes the information needs of down stream end users, including local governments and property owners. As discussed in the wetland comparison, none of the Federal agencies involved designed programs to facilitate the parcel-based regulatory program needs of the state of Maryland. The NRI example shows both sides of this issue. The statistical design to serve regional and national policy needs precludes local analysis, while the data set necessarily integrates site-specific, environmental, ownership, and program attributes. Although EPA's EMAP was called on initially to address local as well as regional and national scales, it quickly abandoned this focus. The new "community" emphasis is more pragmatic, but the agency's traditional reliance on the facility scale and avoidance of parcel information issues will be a partial barrier to this transition. EPA has just begun to recognize the centrality of the parcel in its "brownfield" efforts, working locally with communities to reutilize abandoned contaminated industrial sites.

## REFERENCES

Carbaugh, L. & R. Marx. "The TIGER System," *Government Information Quarterly,* 7 (3) (1990): 285-306.

Clemence, T. "Historical Perspectives on the Decennial Census," *Government Information Quarterly,* 2 (4) (1985): 355-368.

Committee to Review EPA's EMAP, National Research Council. *Review of the Environmental Monitoring and Assessment Program.* Washington, D.C.: National Academy Press, 1992.

Committee to Review EPA's EMAP, National Research Council. *Review of EPA's Environmental Monitoring and Assessment Program: Overall Evaluation.* Washington, D.C.: National Academy Press, 1995.

Committee to Review EPA's EMAP, National Research Council. *Review of the EPA's Environmental Monitoring and Assessment Program: Surface Waters.* Washington, D.C.: National Academy Press, 1994.

Copeland, J. et al. "EPA Program Monitors U.S. Coastal Environments," *GIS World*: (August, 1994): 44-47.

Courtland, S. "Census Confidentiality: Then and Now," *Government Information Quarterly,* 2 (4) (1985): 407-418.

Harlow, J. "History of Soil Conservation Service National Resources Inventories," in *Natural Resources Inventory Training Manual.* Washington, D.C.: USDA Soil Conservation Service, (1994): 1–10.

Mapping Sciences Committee, National Research Council. *Toward a Coordinated Spatial Data Infrastructure for the Nation.* Washington, D.C.: National Accademy Press, 1993.

Martinko, E. "The Environmental Monitoring and Assessment Program." *EPA Monitor* (February 1993): 1.

Marx, R. *Census Geography.* Alexandria, Va American Chamber of Commerce Researchers Association, 1990.

Maryland State Government Geographic Information Coordinating Committee. *Maryland GIS Resource Guide.* Baltimore, MO: The Committee, 1995.

McClure, C., & P. Hernon, ed. *U.S. Scientific and Technical Information (STI) Policies: Views and Perspectives.* Norwood, NJ, Ablex, 1989.

Nature Conservancy. *Standardized National Vegetation Classification System.* Arlington, VA: Nature Conservancy, 1994.

Shapiro, C. *Coordination and Integration of Wetland Data for Status and Trends and Inventory Estimates: Progress Report.* Washington, D.C.: Federal Geographic Data Committee, 1995.

Sperling, J. "Development and Maintenance of the TIGER Database: Experiences in Spatial Data Sharing at the U.S. Bureau of the Census," in *Sharing Geographic Information*, edited by H. Onsrud and G. Rushton. New Brunswick, NJ: Rutgers Center for Urban Policy Research, 1995, pp. 377-395.

Stone, R. "EPA Streamlines Troubled National Ecological Survey.," *Science* 268 (June 9, 1995): 1427–1428.

U.S. Bureau of the Census. Maps and More: Your Guide to Census Bureau Geography. Washington, D.C.: The Agency, 1992.

U.S. Department of Agriculture (1990). *Streamlining the National Resources Inventory Process—Executive Summary.* Washington, D.C.: The Agency, 1990.

U.S. Environmental Protection Agency. *Environmental Goals for America: With Milestones for 2005—Draft for Government Agency Review.* Washington, D.C.: The Agency, 1996.

# 10

## Conclusion

### NEPA'S SUBSTANTIVE INFORMATION FAILURE

*I*n the introductory chapter, NEPA was presented as the turning point in both recognizing the centrality of environmental quality in our nation's health and economic security, and opening up the traditionally closed decision making processes to a more pluralistic dialog. At first, many analysts and agencies dismissed NEPA as a purely symbolic policy. Following a number of high-profile court cases that overturned decisions that had not complied with impact statement requirements, NEPA was taken seriously and bureaucratically internalized.

The successes of NEPA are pervasive. Environmentally insensitive projects now frequently do not proceed beyond the conceptual stage. Other controversial projects go through long litigation and, if implemented, are highly transformed by the negotiations. Although there is still considerable debate over the effectiveness of NEPA, there is a clear consensus that the law's strength is primarily procedural not substantive. Reports must be prepared, parties informed, and feedback processed and "considered" prior to a final decision. The decision tradeoffs themselves do not have to follow any particular rationale; there are no required environmental performance objectives. NEPA is an administrative process not a regulatory program.

Three criticisms of NEPA and the impact statement approach directly relate to its failure to generate deep information: bias, systems ignorance, and one-stop processing. From the outset, it was understood that the impact statements would be biased because the authors are primarily the project proponents (Sax, 1973). Two key components of a statement are the identification and assessment of alternatives. The 60–90 days typically allowed for comments on a draft is often not a feasible period for substantive review. Impact statements are highly technical, large, hard-copy documents. The language and format has evolved to address potential needs in litigation, not to facilitate communications.

Agencies, NGOs, business, and citizens face significant difficulties obtaining, handling, and analyzing their contents. To date, no dissemination policy has emerged to utilize new electronic technologies such as networks and CDs. Impact statements are stand-alone products, unconnected to their underlying databases, GISs, and mathematical models. Without direct access to these resources, review and comment is often superficial. In Chapter 5, we examined the rapid evolution in information dissemination, such as the National Spatial Data Infrastructure initiative.

Digital data distribution is a necessary but insufficient need for NEPA. To be truly participatory, it must also evolve to include "distributed modelling." Stakeholders need the capability to input their own site information, and to develop and assess their own alternatives, as well as analyze the internal assumptions, sensitivities, and accuracies of the project proponents.

Over a decade ago, O'Hare, a major consultant commenting on the prolonged gridlock associated with major impact controversies, stated:

> If people are to bargain successfully over the distribution of benefits from a facility they must know... how it will affect the natural and human environments... we try to generate this type of information through the environmental impact statement. (O'Hare et al., 1983, p. 99)

After reviewing the stakeholders' mistrust of agency analyses and the often wasteful surplus of data generated in anticipation of debate, he proposed radically altering the process: "[we need] to provide each party with a means to obtain his own information." (p. 99)

Distributed data and distributed modelling meet this objective. A pilot effort for this concept exists in the water resources community (Haslam & Wright, 1995). It would take considerable additional development to make such a system functional for a broad set of downstream end users. The National Center for Environmental Decision Making is researching this topic.

The second major deep information critique of the NEPA process is its frequent avoidance of hierarchial systems issues. This critique is similar to the disjointed incrementalism discussion regarding permits and environmental regulations in Chapter 7. Three prime facets of this myopia are the continuing dearth of cumulative impact analyses, the absence of receiving system risk assessments, and the lack of post-decision monitoring.

Underlying the above critiques is a fourth. Although the impact statement approach is a well-articulated administrative process, it never evolved into an integrated deep information system. There are no standards for the data, maps, or models used in impact statements. Since this information is communicated in hard-copy format, there is no potential for integrating information from multiple statements.

In a metropolitan area or major natural resource region, there may have been hundreds of Federal and state impact studies conducted since 1970. These stand-alone reports consist of a diverse range of secondary data and new information created for the particular study. When the review is completed, the report is archived. This approach is not designed to promote community learning over time, wherein each statement contributes new transactional updates to a shared knowledge base. The one-stop approach of the impact statement is highly wasteful of scarce environmental information expenditures, and the lack of a managed information framework condones the mediocre predictive quality that typifies many analyses (Culhane et al., 1985).

## NATIONAL INSTITUTE FOR THE ENVIRONMENT

Most information policies are promulgated as overlays on existing administrative structures. As discussed briefly in Chapter 5, the newer information technologies are providing a major impetus to break down the traditional specialized hierarchial structures of bureaucracies. These are being replaced with more flexible organizations supported by distributed enterprise data models. In some instances, information policy has led to the creation of new agencies or bureaus. The creation of OIRA in OMB has played a pivotal role in shaping Federal policy and programs. Many states have developed new administrative structures to play a similar role both for general information resources and the explosive field of GIS (Warnecke, 1995).

From the previous chapter it was apparent that the nation is facing a growing crisis in the arena of environmental scientific information. Balkanized agency responsibilities, increased end-user demands, and radically diminished budgets are all fueling this problem. Environmental knowledge-building requires integrative, longitudinal, hierarchial data generation, analysis, and dissemination. It is necessarily a highly distributed multi-actor process that currently lacks a coordinating context.

One major proposal to bridge this gap is to form a National Institute for the Environment (NIE) that would be modelled on other existing Federal research institutes. The Committee on the NIE:

> finds that the best, simplest, and most cost-effective way to achieve the four goals (increase understanding, assist decision makers, expand information access, and educate), is to establish *a single agency*... with the central mission of improving the scientific basis for making decisions about environmental issues.... Multiple-agency approaches can bring about positive changes in the current system, but cannot achieve the unique and comprehensive goals of the initiative proposed. (Committee for the National Institute for the Environment, 1994, p. 125)

The proposal calls for the creation of three linked research directorates: Environmental Resources; Environmental Systems; and Environmental Sustainability. It states, "a fundamentally new approach is needed to manage environmental information so that it is credible and easily available to all" (p. 132). This would be the function of a National Library for the Environment (NLE). The NLE would develop a strategy for harnessing our disjointed, rapidly expanding ability to generate environmental data into a standardized system which provides one-point access, and data quality metadata for end users with "diverse expertise and varied sensitivities."

The NIE proposal has gained considerable Congressional, academic, and industrial support. It is difficult to argue with its logic. One inherent problem with the NIE, however, is its elitist perspective that knowledge building is top-down effort by researchers. As discussed in Chapters 6, 7, and 8, much of our environmental data and understanding is generated at the site level through the regulatory processes and day-to-day stewardship of owner and operators. This data is inherently messy and has numerous quality control problems. It also is the richest continuing source of in situ monitoring and environmental learning by the responsible parties. Although the EMAP and other examples have demonstrated the political impracticality of combining research and regulatory activities, permit and cadastral data will be essential to comprehensive systems knowledge building.

## DEEP PERSPECTIVES

Some dimensions of deep information are evolving rapidly as integral components of the larger moves to improve governmental IRM. Examples include locator systems and the National Spatial Data Infrastructure. These undertakings facilitate integration and dissemination of environmental information, particularly at regional and national scales, but generally ignore the deep ecology principle that the entire economy and social structure are enfolded in the physical environment.

To date, Federal and state information policies have failed to acknowledge the need to integrate fully land ownership records with environmental information. The cadastre is still seen as solely a state property taxation policy arena, while pollution control and natural resource protection has been primarily viewed as a Federal role. The latter is about to shift dramatically as authority is significantly transferred to the states.

Constitutionally, Federal agencies such as EPA and the Department of Agriculture can't dictate state and local LIS activities, but they can develop regulatory and provision program information reporting policies, which promote the use of standardized cadastral key attributes, such as PINs. The decentralization of environmental management functions from Washington to the states and locales will heighten the need for more robust information on which to build knowledge of environmental process change.

Deep information requires the proactive synthesis of technology and policy changes, but these actions alone are insufficient to generate sustainability. As discussed in Chapter 4, the traditional map paradigms of mechanistic thematic and contour portrayals are obsolete. Deep maps which reflect both system dynamics and GIS have not yet emerged, due in part to cultural inertia and in part to the need for object-oriented data systems, which will facilitate smart map elements.

In Chapter 3, we left Greenetowne with some serious endemic challenges. Even with the assistance of state and national programs, the community is faced with locally developing a pervasive understanding of its linkages to its environmental context, and initiating a spectrum of fundamental changes in its personal, corporate, and public sector values and activity patterns. Early in the century, Dewey had argued that communities have the inherent capacity to make such decisions if they are well informed. Lee, based on his work in the Pacific Northwest, has concluded that there are no single environmental decisions. All actions are experiments in process that require structured feedback and periodic midcourse corrections. Ridley and Low argue that this value shift can only take place when we see the clear linkages between environmental health and economic wellbeing (Ridley & Low, 1993).

Beyond NEPA's primary focus on the establishment of formal agency procedures are a number of symbolic but wise pronouncements. Among these is:

> The Congress recognizes that each person should enjoy a healthful environment and that each person has a responsibility to contribute to the preservation of the environment (NEPA Sec. 4331 (c))

Jane Jacobs would agree.

## REFERENCES

Committee for the National Institute for the Environment. "A Proposal to Create a National Institute for the Environment (NIE)," *The Environmental Professional,* 16 (2) (1994): 93–175.

Culhane, P. et al. *Forecasts and Environmental Decision Making: The Contents and Accuracy of Environmental Impact Statements.* Evanston, IL: Center for Urban Affairs and Policy Research, Northwestern University, 1985.

Haslam, E., & J. Wright. "Indiana WETnet: A Virtual Water Resource," *Water Resources Update,* 99 (1995): 3–11.

O'Hare, M. et al. *Facility Siting and Public Opposition.* New York, Van Nostrand Reinhold, 1983.

Ridley, M., & B. Low. "Can Selfishness Save the Environment?" *Atlantic Monthly* (Sept. 1993): 76–86.

Sax, J. "The (Unhappy) Truth about NEPA," *Oklahoma Law Review* 26 (1973): pp. 239–48.

Warnecke, L. *GIS Institutionalization in the 50 States.* Santa Barbara, CA: National Center for Geographic Information and Analysis, 1995.

# Appendix

........................

## ACRONYMS

APA Administrative Procedures Act
CAD Computer Assisted Design
CERCLA Comprehensive Environmental Response, Compensation, and Liability Act
COE (U.S. Army) Corps of Engineers
DEC (NY) Department of Environmental Conservation
DEM Digital Elevation Model
DFD Data Flow Diagram
DFIRM Digital Flood Insurance Rate Map
DLG Digital Line Graph
DO Dissolved Oxygen
EIS Environmental Impact Statement
EMAP Environmental Monitoring and Assessment Program
EPA Environmental Protection Agency
EPCRA Emergency Planning and Community Right-to-Know Act
FEMA Federal Emergency Management Agency
FGDC Federal Geographic Data Committee
FIPS Federal Information Processing Standard
FIRM Flood Insurance Rate Map
FOIA Freedom of Information Act
FWS Fish and Wildlife Service
GIS Geographic Information System
IDOC Illinois Department of Conservation
IDOT Illinois Department of Transportation
IRM Information Resource Management
ISIS Illinois Stream Information System
LEPC Local Emergency Planning Committee
LIS Land Information Systems
LUNR (NY) Land Use and Natural Resources inventory
MPLIS Multipurpose Land Information System
MSDS Material Safety Data Sheet
NBS National Biological Survey
NCRS Natural Resources Conservation Service

NEPA National Environmental Policy Act
NGO Non-Governmental Organization
NRI National Resources Inventory
NOAA National Oceanographic and Atmospheric Administration
NSDI National Spatial Data Infrastructure
NWI National Wetlands Inventory
OIRA Office of Information and Regulatory Affairs
OMB Office of Management and Budget
OSHA Occupational Safety and Health Administration
PIN Parcel Identification Number
PPA Pollution Prevention Act
RCIS (NY) Regulatory Compliance Information System
SCS Soil Conservation Service
SERC State Emergency Response Committee
SIC Standard Industrial Code
TIGER Topologically Integrated Geographic Encoding and Referencing system
TRI Toxic Release Inventory
USDA U.S. Department of Agriculture
USGS U.S. Geological Survey
UTM Universal Transverse Mercator
ABA American Bar Association
BOD Biochemical Oxygen Demand
FINDS Facility Index System
BMP Best Management Practices
GILS Government Information Locator System
GPS Global Positioning System
MIS Management Information System
NPDES National Pollution Discharge Elimination System
NIE National Institute of Environment
PLS Public Land Survey
RCRA Resource Conservation and Recovery Act
SARA Superfund Ammendments and Reauthorization Act
USC US Code
PPM Parts per million

# Biographical Note on Author

...........................

John Felleman is a Professor of Environmental Studies at the College of Environmental Science and Forestry, State University of New York, Syracuse, NY 13210 (felleman@mailbox.syr.edu). he has a bachelors and Masters of Civil Engineering from Cornell University and a Doctor of Public Administration from New York University. He has worked as a planner and environmental analyst in government and consulting. Current teaching and research includes: predictive modelling, environmental decision making, environmental information policy, and visualization.

# Author Index

# Subject Index

...........................